Fundamental Theories of Physics

Volume 202

The international monograph series "Fundamental Theories of Physics" aims to stretch the boundaries of mainstream physics by clarifying and developing the theoretical and conceptual framework of physics and by applying it to a wide range of interdisciplinary scientific fields. Original contributions in well-established fields such as Quantum Physics, Relativity Theory, Cosmology, Quantum Field Theory, Statistical Mechanics and Nonlinear Dynamics are welcome. The series also provides a forum for non-conventional approaches to these fields. Publications should present new and promising ideas, with prospects for their further development, and carefully show how they connect to conventional views of the topic. Although the aim of this series is to go beyond established mainstream physics, a high profile and open-minded Editorial Board will evaluate all contributions carefully to ensure a high scientific standard.

More information about this series at http://www.springer.com/series/6001

Albert Petrov

Quantum Superfield Supersymmetry

 Springer

Albert Petrov ⓘ
Departamento de Fisica
Universidade Federal da Paraiba
João Pessoa, Brazil

ISSN 0168-1222   ISSN 2365-6425 (electronic)
Fundamental Theories of Physics
ISBN 978-3-030-68135-7   ISBN 978-3-030-68136-4 (eBook)
https://doi.org/10.1007/978-3-030-68136-4

This Springer imprint is published by the registered company Springer Nature Switzerland AG
The registered company address is: Gewerbestrasse 11, 6330 Cham, Switzerland

Preface

This review represents itself as a collection of lecture notes on superfield supersymmetry based on lectures given at Instituto de Física, Universidade de São Paulo (São Paulo), Instituto de Física, Universidade Federal do Rio Grande do Sul (Porto Alegre), and Departamento de Física, Universidade Federal da Paraiba (João Pessoa). The book includes many examples of explicit quantum calculations performed with use of the superfield formalism and is intended for students at levels of undergraduate and graduate studies in quantum field theory and theory of elementary particles and for researchers working in related subjects.

Author is grateful to C. A. S. Almeida, E. A. Asano, L. C. T. Brito, I. L. Buchbinder, M. Cvetic, A. F. Ferrari, F. S. Gama, H. O. Girotti, M. Gomes, S. M. Kuzenko, A. C. Lehum, R. V. Maluf, J. R. Nascimento, P. Porfirio, A. A. Ribeiro, V. O. Rivelles, A. J. da Silva, and E. O. Silva for fruitful collaboration and interesting discussions. The work has been partially supported by CNPq.

João Pessoa, Brazil   Albert Petrov

Contents

Chapter 1
Introduction

*We prove, once and for all, that people who don't use
superspace are really out of it.*

"Stuperspace"

The idea of supersymmetry is now considered as one of the basic concepts of theoretical high energy physics (see e.g. [1]). The supersymmetry, being a fundamental symmetry allowing to relate bosons and fermions, provides possibilities to construct theories with much better renormalization properties since some bosonic and fermionic divergent contributions in supersymmetric theories cancel each other. Moreover, there are essentially finite four-dimensional supersymmetry theories without higher derivatives, e.g. $\mathcal{N} = 4$ super-Yang-Mills (SYM) theory (the detailed discussion of the finiteness of this theory is presented in [2]), and, probably, $\mathcal{N} = 8$ supergravity [3]. In the three-dimensional case, $\mathcal{N} = 6$ and $\mathcal{N} = 8$ supersymmetric Chern-Simons theories are known to be finite [4]. Due to this essential improvement of renormalization properties, there is a common expectation that the expected unified theory of all fundamental interactions must be supersymmetric (see e.g. [5] and references therein).

The concept of supersymmetry was introduced in known papers by Volkov and Akulov [6] and Golfand and Lichtman [7] in early 70s. It got a further advance in [8, 9] (the history of arising and development of the concept of the supersymmetry, including biographical aspects, can be found in the book [10]). The essential breakthrough in supersymmetric field theory was achieved with introducing the idea of a superfield [11] (see also [12, 13]). The reason for it consists in the fact that the superfield formulation, first, allows to maintain manifest supersymmetric covariance at all steps of calculations, with all calculations are performed in a very compact manner (indeed, any supergraph corresponds to a set of the Feynman diagrams for component fields), second, automatically takes into account the famous "miraculous cancellations" of ultraviolet divergences responsible for an essential improvement of the renormalization behavior of supersymmetric field theories [14]. Moreover, in the context of the noncommutative field theory it turns out to be that the supersym-

© The Author(s), under exclusive license to Springer Nature Switzerland AG 2021
A. Petrov, *Quantum Superfield Supersymmetry*, Fundamental Theories of Physics 202,
https://doi.org/10.1007/978-3-030-68136-4_1

metry allows for the cancellation not only of ultraviolet divergences, as usual, but also of dangerous infrared divergences arising due to the UV/IR mixing mechanism (see e.g. [15]). Further, the concept of the extended supersymmetry has been introduced, and various versions of an extended supersymmetric formalism have been elaborated. Within this lecture course, nevertheless, we concentrate on the $\mathcal{N} = 1$ superfield formalism which is known as one of the most universal tools for studying the supersymmetric field theories.

Now, let us briefly describe the main steps in development of the superfield methodology. The first example of the successful application of the superfield concept was the model proposed by Wess and Zumino in their seminal paper [11] where the simplest superfield model, further called the Wess-Zumino model, has been formulated. Further, in [9] they introduced the superfield gauge model, that is, the SYM theory. Intensive studies of different issues related quantum aspects of these theories began to be carried out. One of the most important results is the finiteness of $\mathcal{N} = 4$ SYM theory that was proved in [16] (for discussions of finiteness of the supersymmetric gauge theories see also [17]), which implied in a strongest interest to the supergauge theories. Then, the superfield supergravity was formulated [18]. In 1984, the first consistent methodology possessing an explicit extended ($\mathcal{N} = 2$) supersymmetry has been developed, that is, the harmonic superspace [19, 20]. At the same year, the superfield approach has been successfully applied to the superstring theory [21], which emphasized the importance of this methodology within the string context. A bit earlier, the superfield formulation has been developed for the supersymmetric theories in a three-dimensional space-time [22, 23], which further manifested itself as a very convenient laboratory for studies of different issues related to the supersymmetry. It should be noted that actually, the interest to three-dimensional field theories, besides of their simplicity and better renormalization properties, is driven by studies of graphene. In our context, it worth mentioning that the superfield supersymmetric model for graphene has been formulated as well [24].

A new epoch for studies of the superfield theories began in 1991 when, in the papers [25, 26], the chiral quantum contributions to the effective action were discussed for the first time. These papers called an attention to the superfield methodology for evaluating the effective potential whose development has been carried out in a series of works initiated by the paper [27], where this methodology has been successfully applied to the Wess-Zumino model. Further, the success of the paper [28] strongly increased the interest to various issues, both perturbative and non-perturbative ones, related to supergauge theories, especially those ones possessing the extended supersymmetry.

Among the most successful applications of the superfield methodology, also its use for the noncommutative supersymmetric field theories deserves to be mentioned. The famous paper [29] established the fact that, since the space-time noncommutativity does not affect the anticommuting (Grassmannian) coordinates of superfields, the superfield methodology can be used as a very powerful instrument to deal with the famous problem of the UV/IR mixing [30] known for generating the new infrared divergences which are able to break the perturbative expansion. It was shown first in [29] that the supersymmetric extension of field theories, implying in improving the ultraviolet behavior, also allows to remove the dangerous infrared divergences arising due to the UV/IR mechanism. Further, the non(anti)commutativity has been introduced also to the fermionic sector [31], however, actually it implies in modifying the corresponding superfield action by a small number of additive terms proportional to lower degrees of the non(anti)commutativity parameter. In all other aspects, there is no essential difference between superfield theories formulated on the base of the fermionic noncommutativity and the usual ones, and in these lecture notes, the fermionic non(anti)commutativity is discussed only very briefly.

The $\mathcal{N} = 1$ superfield methodology in supersymmetric quantum field theory, being an universal tool for a great number of supersymmetric models, is a main topic of these lectures. We consider the superfield description both of three- and four-dimensional supersymmetric field theories, including the noncommutative generalization for some cases. In the chapter devoted to three-dimensional theories we use the notations and conventions introduced in [23], while in the chapter devoted to four-dimensional theories—those ones introduced in [32, 33].

Within this review, we are going not only to describe the superfield formalism in three- and four-dimensional space-times and give the superfield formulation for the most important examples of the supersymmetric field theories, but also to consider in details many examples of calculating loop corrections within the superfield description. Our review is focused, principally, in evaluating a superfield effective action in various supersymmetric field theories. Because of the restricted value of the lecture course, we do not discuss here the superstring and supergravity issues, for which we recommend [1, 33] respectively.

The structure of this review looks like follows. In the Chap. 2, we give a basic description of an effective action and the loop expansion. In the Chap. 3 we discuss the superfield formalism, the supergraph technique and methods of calculating the superfield effective action in the three-dimensional space-time. In the Chap. 4, the four-dimensional superfield methodology is described, and many examples of quantum calculations are given. In the Chap. 5, we present the introduction to the problem of a supersymmetry breaking. Finally, in the Summary we discuss the perspectives and applications of supersymmetry.

Chapter 2
Effective Action and Loop Expansion: General Formulation

Before entering the discussion of superfield theories, let us give the general definition of the effective action and describe methods for its calculation. An effective action is a key object of quantum field theory. Studying of an effective action allows to investigate problems of vacuum stability, Green functions, spontaneous symmetry breaking, anomalies and many other issues. In this chapter we present the description of its general structure and manners of its calculation, which in the next chapters will be applied to supersymmetric field theories. We follow the methodology described in [34, 35].

An effective action (in particular, in a superfield theory) is defined, as usual, as a generating functional of one-particle-irreducible Green functions. It is obtained as a Legendre transform for the generating functional of connected Green functions:

$$\Gamma[\Phi] = W[J] - \int dz J(z)\Phi(z). \tag{2.1}$$

Here $\Gamma[\Phi]$ is an effective action, dz denotes an integral over the corresponding space or its subspace (in some cases, to abbreviate the notations we omit the sign of the integral together with the coordinate dependence of the fields, i.e. $J\Phi \equiv \int dz J(z)\Phi(z)$), $J(z)$ is the classical source, $\Phi(z) = \frac{\delta W[J]}{\delta J(z)}$ is a so called mean (super)field, or, as in the same, a background (super)field, which is an essentially classical object (in principle, one can consider a set of background (super)fields as well, so, in general, Φ is a column vector); in terms of the functional integral one has

$$\Phi = \frac{\int D\phi\, \phi\, e^{iS[\phi]+i\phi J}}{\int D\phi\, e^{iS[\phi]+i\phi J}},$$

where $S[\phi]$ is the classical action. This is exactly the definition of the vacuum expected value (v.e.v.) in the path integral formalism, and $W[J] = \frac{1}{i}\log Z[J]$ is a generating functional of the connected Green functions. It is easy to see that the $\Gamma[\Phi]$ satisfies the equation

© The Author(s), under exclusive license to Springer Nature Switzerland AG 2021
A. Petrov, *Quantum Superfield Supersymmetry*, Fundamental Theories of Physics 202,
https://doi.org/10.1007/978-3-030-68136-4_2

$$\frac{\delta \Gamma[\Phi]}{\delta \Phi(x)} = -J(x).$$

The effective action can be expressed in the form of the path integral [35]:

$$e^{\frac{i}{\hbar}\Gamma[\Phi]} = \int D\phi \, e^{\frac{i}{\hbar}(S[\phi]+\phi J - \Phi J)}. \tag{2.2}$$

Here $S[\phi]$ is the classical action of the corresponding theory. Note that ϕ is an integration variable, and Φ is a function of a classical source J which does not depend on ϕ. We introduced the Planck constant $\hbar$ by dimensional reasons (indeed, the Planck constant has a dimension of the action) and in order to obtain the loop expansion. We note that, to distinguish quantum effects from the classical ones, we treat $\hbar$ as a small parameter, thus, the effective action is a power series in $\hbar$. To calculate this integral we make change of variables of integration:

$$\phi \rightarrow \Phi + \sqrt{\hbar}\phi.$$

If the theory describes several fields we can unite them into a column vector, and all consideration is quite analogous. The integral (2.2) after this change takes the form

$$e^{\frac{i}{\hbar}\Gamma[\Phi]} = \int D\phi \, e^{\frac{i}{\hbar}(S[\Phi+\sqrt{\hbar}\phi]+\sqrt{\hbar}\phi J)}. \tag{2.3}$$

Here, our aim consists in the obtaining the expansion of $\Gamma[\Phi]$ in power series in $\hbar$ following the approach described in [35].

To start our calculation, we expand the factor in the exponential of the above expression into power series in $\hbar$:

$$\frac{i}{\hbar}S[\Phi+\sqrt{\hbar}\phi] + \frac{i}{\sqrt{\hbar}}\phi J = \frac{i}{\hbar}\Big(S[\Phi] + (S'[\Phi]+J)\sqrt{\hbar}\phi + \frac{\hbar}{2}S''[\Phi]\phi^2 + \dots +$$
$$+ \frac{\hbar^{n/2}}{n!}S^{(n)}[\Phi]\phi^n + \dots \Big). \tag{2.4}$$

Here $S^{(n)}[\Phi]$ denotes n-th variational derivative of the classical action with respect to Φ (integration over the corresponding space is assumed). This expansion can be substituted into (2.3). We then introduce the quantum part of the effective action, $\bar{\Gamma}[\Phi] = \Gamma[\Phi] - S[\Phi]$, which can be expanded into power series in $\hbar$, beginning from the first order: $\bar{\Gamma} = \sum\limits_{n=1}^{\infty} \hbar^n \Gamma^{(n)}$. As a result we have

$$e^{\frac{i}{\hbar}\bar{\Gamma}[\Phi]} = \int D\phi \, \exp\Big[\frac{i}{\hbar}\Big(\sqrt{\hbar}(S'[\Phi]+J)\phi + \frac{\hbar}{2}S''[\Phi]\phi^2 + \dots +$$
$$+ \frac{\hbar^{n/2}}{n!}S^{(n)}[\Phi]\phi^n + \dots \Big)\Big]. \tag{2.5}$$

It is clear that the first block, $\frac{i}{\sqrt{\hbar}}(S'[\Phi] + J)\phi$, can lead only to one-particle-reducible Feynman (super)graphs since its contribution with one quantum field ϕ can form only one propagator when the Feynman diagrams are constructed, thus it does not contribute to the effective action. Hence we can omit this term. Then we can expand the exponent into power series in $\hbar$:

$$e^{\frac{i}{\hbar}\bar{\Gamma}[\Phi]} = \int D\phi \, e^{\frac{i}{2}S''[\Phi]\phi^2} \left(1 + \frac{i\sqrt{\hbar}}{3!}S^{(3)}[\Phi]\phi^3 + \frac{i\hbar}{4!}S^{(4)}[\Phi]\phi^4 + \right.$$
$$\left. +(1/2)\left(\frac{i\sqrt{\hbar}}{3!}\right)^2 (S^{(3)}[\Phi]\phi^3)^2 + \dots\right). \tag{2.6}$$

At the same time, after substituting the expansion of $\bar{\Gamma}$ in the left-hand side of (2.5) into power series in $\hbar$, we get:

$$\exp\left(\frac{i}{\hbar}\bar{\Gamma}[\Phi]\right) = e^{i(\Gamma^{(1)}[\Phi] + \hbar\Gamma^{(2)}[\Phi] + \dots)} = e^{i\Gamma^{(1)}[\Phi]}(1 + i\hbar\Gamma^{(2)}[\Phi] + \dots).$$

Substituting this expansion into (2.6) and comparing equal powers of $\hbar$ we see that any correction $\Gamma^{(n)}$ corresponds to a some functional integral. For example, the correction of the first order in $\hbar$ (i.e., as we will show further, the one-loop correction) is defined from the paradigmatic integral

$$\exp(i\Gamma^{(1)}[\Phi]) = \int D\phi \, e^{\frac{i}{2}S''[\Phi]\phi^2}, \tag{2.7}$$

and the correction of the second order in $\hbar$—from the relation

$$\Gamma^{(2)}[\Phi] = \frac{1}{i}\frac{\int D\phi \, e^{(\frac{i}{2}S''[\Phi]\phi^2)}\left(\frac{i}{4!}S^{(4)}[\Phi]\phi^4 - \frac{1}{2(3!)^2}(S^{(3)}[\Phi]\phi^3)^2\right)}{\int D\phi \exp(\frac{i}{2}S''[\Phi]\phi^2)}. \tag{2.8}$$

Here, as usual, integration over coordinates in expressions of the form $S^{(n)}[\Phi]\phi^n$ is assumed. The denominator of this expression serves to eliminate the one-particle-reducible contributions.

We can see that:

(i) All fractional degrees of $\hbar$ like $\hbar^{n+1/2}$, with integer n, vanish since they accompany integrals like $\int D\phi\phi^{2n+1} \exp(\frac{i}{2}S''[\Phi]\phi^2)$. By the symmetry reasons such an expression is equal to zero.

(ii) All terms beyond the first order in $\hbar$ are expressed in the form of some functional integrals.

(iii) The one-loop correction (2.7) can be expressed in the form of a functional determinant since

$$\int D\phi \exp(\frac{i}{2}S''[\Phi]\phi^2) = \mathrm{Det}^{-1/2}S''[\Phi], \tag{2.9}$$

which leads to

$$\Gamma^{(1)} = \frac{i}{2}\mathrm{Tr}\log S''[\Phi]. \tag{2.10}$$

The $S''[\Phi] \equiv \Delta$ is a some operator. In most typical cases without higher derivatives it has the form $\Delta = \Box + (\ldots)$, with dots are for terms of lower orders in derivatives, for example, a mass (in particular, a background dependent one). We can express the one-loop effective action in terms of the functional (super)trace

$$\Gamma^{(1)} = \frac{i}{2}\mathrm{Tr}\int_0^\infty \frac{ds}{s}e^{is\Delta}. \tag{2.11}$$

This expression is called the Schwinger representation for the one-loop effective action. The sign Tr denotes both a matrix trace tr (if Δ possesses matrix indices) and a functional trace, i.e.

$$\mathrm{Tr}\,e^{is\Delta} = \mathrm{tr}\int d^n z_1 d^n z_2 \delta^n(z_1 - z_2)e^{is\Delta}\delta^n(z_1 - z_2).$$

Here n is a dimension of the corresponding (super)space. The calculation of $e^{is\Delta}$ in field theories is carried out with use of a special procedure called Schwinger-De Witt method or the proper time method [36]. The realization of this method in superfield theories will be discussed further.

Let us consider higher orders in the Planck constant. From (2.6) it is easy to see that all loop corrections beyond the first order in $\hbar$ have the form of some functional integrals, i.e. they look like

$$\int D\phi \exp(\frac{i}{2}S''[\Phi]\phi^2)\prod_n (S^{(n)}[\Phi]\phi^n). \tag{2.12}$$

Such integrals can be calculated in the way analogous to the standard perturbative methodology. We can use the identity

$$\int D\phi \phi^n e^{i\frac{1}{2}\phi\Delta\phi} = \left(\frac{1}{i}\frac{\delta}{\delta j}\right)^n \int D\phi e^{i(\frac{1}{2}\phi\Delta\phi + j\phi)}|_{j=0}, \tag{2.13}$$

which allows to develop the Feynman diagram technique in which the role of vertices is played by $\frac{S^{(n)}[\Phi]\phi^n}{n!}$, and the role of propagators—by $i\Delta^{-1}$. However, since the operator $\Delta = S''[\Phi]$, in general, is background dependent (see above) we arrive at background dependent propagators $< \phi(z_1)\phi(z_2) >= i\Delta^{-1}\delta(z_1 - z_2)$. In superfield theories, these propagators, typically, can be found exactly only in some special cases, the most important of them are, first, the case of background superfields constant in space-time, second, the case of chiral background superfields only. Further we consider some examples.

Let us turn again to (2.6). We see that each quantum superfield is accompanied by $\hbar^{-1/2}$, and each vertex—by $\hbar^{-1}$ (which provides a $\hbar^{n/2-1} S^{(n)}[\Phi]\phi^n$ form of the vertex). An arbitrary (super)graph with P propagators and V vertices contains $2P$ quantum superfields (indeed, each propagator is formed by contraction of two superfields). Therefore if this (super)graph contain vertices $S^{(n_1)}[\Phi]\phi^{n_1}, S^{(n_2)}[\Phi]\phi^{n_2}, \ldots, S^{(n_V)}[\Phi]\phi^{n_V}$, its power in $\hbar$ is $\sum_{i=1}^{V}(\frac{n_i}{2} - 1) = \frac{1}{2}\sum_{i=1}^{V} n_i - V$. However, $\sum_{i=1}^{V} n_i$ is just the number of quantum fields associated with all vertices which is equal to $2P$. Therefore the contribution described by this (super)graph has the power of $\hbar$ equal to $P - V = L - 1$, with L is number of loops. But any expression of the form (2.6), by the definition, is a contribution to $\frac{\Gamma}{\hbar}$, hence a contribution from L-loop (super)graph to Γ is proportional to $\hbar^L$. Hence we proved that the order in $\hbar$ from an arbitrary (super)graph is just the number of loops in it, and the expansion in powers of $\hbar$ is called the loop expansion. As a result we see that loop corrections can be calculated on the base of a special (super)field technique.

In the next chapters of this review we will apply this methodology through evaluating the effective action in the superfield formalism both in three- and four-dimensional superspaces, calculating explicitly one- and two-loop contributions to the effective actions of different superfield theory models.

Chapter 3
Superfield Description of Three-Dimensional Supersymmetric Theories

In this chapter we describe the superfield formalism for three-dimensional supersymmetric field theory models. Afterwards, we describe various approaches to calculate quantum corrections in three-dimensional superfield theories, including the noncommutative ones. These approaches are illustrated with different examples discussed in details.

3.1 Definitions and Conventions

The basic concept of supersymmetric field theory consists in the existence of some essentially new symmetry transformations with a fermionic parameter which mix fermionic and bosonic dynamical variables of the theory. To provide a nontrivial connection of the supersymmetry transformations with usual Poincaré transformations, we also suggest that the anticommutator of two supersymmetry generators differs from zero. In the most used, and simplest, versions of the supersymmetry algebra, including the versions considered in this book, this anticommutator is proportional to a space-time translation.

In this chapter we follow conventions of [23]: in the three-dimensional space-time we choose the Minkowski metric of the form $\eta_{mn} = \text{diag}(- + +)$, and the Dirac matrices are the 2×2 matrices whose explicit form is: $(\gamma^0)^\alpha{}_\beta = -i\sigma^2, (\gamma^1)^\alpha{}_\beta = \sigma^1, (\gamma^2)^\alpha{}_\beta = \sigma^3$, with $\{\gamma^m, \gamma^n\} = 2\eta^{mn}$. We use bispinor notations based on converting any vector index into two spinor indices by the rule $A^m \to A^{\alpha\beta} = A^m (\gamma_m)^{\alpha\beta}$. We note that the Dirac matrices with two lower indices are $(\gamma^m)_{\alpha\beta} = (-\mathbf{1}_2, -\sigma^3, \sigma^1)$ are symmetric, hence all vectors (in particular, coordinates x^m and corresponding derivatives ∂_m) can be mapped into *symmetric* bispinors. To raise and lower the spinor indices we use the Hermitian Levi-Civita-like C symbol: $C_{\alpha\beta} = -i\epsilon_{\alpha\beta} = \begin{pmatrix} 0 & -i \\ i & 0 \end{pmatrix} = -C^{\alpha\beta}$, with $\psi^\alpha = C^{\alpha\beta}\psi_\beta, \psi_\alpha = \psi^\beta C_{\beta\alpha}$ ("north-western" convention). Also, one has $C^{\alpha\beta}C_{\beta\gamma} = -\delta^\alpha_\gamma$, and $C^{\alpha\beta}C_{\alpha\beta} = 2$. We define $\psi^2 = \frac{1}{2}\psi^\alpha\psi_\alpha$, and use

© The Author(s), under exclusive license to Springer Nature Switzerland AG 2021
A. Petrov, *Quantum Superfield Supersymmetry*, Fundamental Theories of Physics 202,
https://doi.org/10.1007/978-3-030-68136-4_3

the identity $A_{[\alpha} B_{\beta]} = -C_{\alpha\beta} A^\gamma B_\gamma$ following from the properties of the irreducible representations of the Lorentz group, i.e. the (anti)symmetrization will be henceforth defined without the $1/n!$ factor. The reader should note the presence of $\frac{1}{2}$ factor in all "square" contractions like $D^2 = \frac{1}{2} D^\alpha D_\alpha$, $W^2 = \frac{1}{2} W^\alpha W_\alpha$, etc. This is the difference from the conventions for the three-dimensional superspace proposed by Ruiz Ruiz and Nieuwenhuizen in [37].

To formulate the superspace, we start with introducing spinor coordinates θ^α with $\alpha = 1, 2$. These coordinates are transformed under the spinor representation of three-dimensional Lorentz group, i.e. $SO(1, 2)$ group (the spinor representation of the Lorentz group in the three-dimensional space-time is given by the $SL(2, R)$ group). In a general case, there can be $\mathcal{N}$ sets of spinor coordinates $\theta^{i\alpha}$, with $i = 1, \ldots \mathcal{N}$. However, throughout this book, both in three- and four-dimensional cases, we will consider only the $\mathcal{N} = 1$ superspace, i.e. we assume that there is only one set of spinor coordinates and, correspondingly, only one set of supersymmetry generators. The spinor coordinates θ^α, satisfying the Grassmannian anticommutation condition $\{\theta^\alpha, \theta^\beta\} = 0$, together with the usual bosonic coordinates x^m parametrize the superspace.[1]

To develop field theory on superspace we must introduce integration and differentiation on superspace, i.e. with respect to Grassmannian coordinates [38]. We can introduce left ∂_L and right ∂_R derivatives with respect to Grassmannian coordinates as

$$\frac{\partial_L}{\partial \theta^{\alpha_i}} (\theta^{\alpha_1} \theta^{\alpha_{i-1}} \theta^{\alpha_i} \theta^{\alpha_{i+1}} \ldots \theta^{\alpha_n}) = (-1)^{i-1} (\theta^{\alpha_1} \theta^{\alpha_{i-1}} \theta^{\alpha_{i+1}} \ldots \theta^{\alpha_n});$$

$$\frac{\partial_R}{\partial \theta^{\alpha_i}} (\theta^{\alpha_1} \theta^{\alpha_{i-1}} \theta^{\alpha_i} \theta^{\alpha_{i+1}} \ldots \theta^{\alpha_n}) = (-1)^{n-i+1} (\theta^{\alpha_1} \theta^{\alpha_{i-1}} \theta^{\alpha_{i+1}} \ldots \theta^{\alpha_n}). \quad (3.1)$$

Therefore these derivatives differ only by a sign factor. So, it is natural to choose one of them, for example, the left one, and use it henceforth. Thus, we can define the derivatives with respect to usual and Grassmannian coordinates as follows:

$$\partial_\alpha \equiv \frac{\partial}{\partial \theta^\alpha}; \quad \partial_\alpha \theta^\beta = \delta_\alpha^\beta;$$

$$\partial_{\alpha\beta} x^{\gamma\delta} = \delta_{(\alpha}^\gamma \delta_{\beta)}^\delta; \quad \Box = \frac{1}{2} \partial_{\alpha\beta} \partial^{\alpha\beta}. \quad (3.2)$$

Here $A_{(\alpha} B_{\beta)} \equiv \frac{1}{2}(A_\alpha B_\beta + A_\beta B_\alpha)$ is a symmetrized product of spinors A_α and B_β.

The superfield is a (general) function of the superspace coordinates which can be introduced in the form the Taylor series in θ which is finite because of the anticommuting nature of θ^α:

$$f(x, \theta) = f_0(x) + f_1^\alpha(x) \theta_\alpha + f_2(x) \theta^2, \quad (3.3)$$

---

[1] In the Sect. 3.7 we will briefly discuss the simplest deformation of the anticommutation relation, that is, $\{\theta^\alpha, \theta^\beta\} = C^{\alpha\beta}$, in that case the spinor coordinates form the Clifford algebra instead of the Grassmann algebra; this methodology has been originally introduced in [31].

where $\theta^2 = \frac{1}{2}\theta^\alpha\theta_\alpha$. It is clear that if $f_0(x)$, $f_1(x)$, $f_2(x)$ are physical fields, the spin of $f_1^\alpha(x)$ differs by $1/2$ from the spin of $f_0(x)$ and of $f_2(x)$, i.e. $f_1^\alpha(x)$ is a spinor field while $f_0(x)$, $f_2(x)$ are the scalar ones. In a general case, the $f_0(x)$, $f_1(x)$, $f_2(x)$ can carry extra indices being tensors of different ranks, the simplest example is the spinor superfield we discuss further.

The integral over the Grassmannian coordinates is defined as

$$\int d\theta_\alpha \theta^\beta = \delta_\alpha^\beta, \tag{3.4}$$

which is frequently referred as the conclusion that for Grassmann variables, the integration is equivalent to the differentiation. We note that actually it implies that the mass dimensions of θ_α and $d\theta_\alpha$ are not equal but opposite. Actually, for θ_α the dimension is $-1/2$, and for $d\theta_\beta$ or ∂_β, is $1/2$, while for $\partial_{\alpha\beta}$ it is 1 as it must be for the space-time derivative. With using of the definition $d^2\theta = \frac{1}{2}d\theta^\alpha d\theta_\alpha$, we conclude that the integral (3.4) implies

$$\int d^2\theta \, \theta^2 = -1. \tag{3.5}$$

This result allows to define the Grassmannian delta function

$$\delta^2(\theta) = -\theta^2, \tag{3.6}$$

which satisfies the usual property of the delta function

$$\int d^2\theta_1 f(\theta_1)\delta^2(\theta_1 - \theta_2) = f(\theta_2). \tag{3.7}$$

Further, in this chapter we will denote $\delta^2(\theta_1 - \theta_2)$ as δ_{12}. We also will employ the notation $\delta^5(z_1 - z_2) \equiv \delta(z_1 - z_2) = \delta^3(x_1 - x_2)\delta_{12}$. The generators of the supersymmetry transformations, playing the role of translations in the superspace and also called the supercharges, are defined as

$$Q_\alpha = i\partial_\alpha + \theta^\beta \partial_{\beta\alpha}. \tag{3.8}$$

They evidently commute with the generators of "common" bosonic translations $P_{\alpha\beta} = i\partial_{\alpha\beta}$, while the anticommutator of two supercharges is nontrivial:

$$\{Q_\alpha, Q_\beta\} = 2P_{\alpha\beta}. \tag{3.9}$$

We note that the mass dimension of Q_α is equal to $1/2$. The supersymmetry transformation for the given superfield $\Phi(x, \theta)$ with the infinitesimal parameter ϵ^α is defined as

$$\delta\Phi(x, \theta) = -i\epsilon^\alpha Q_\alpha \Phi(x, \theta). \tag{3.10}$$

Projecting this transformation to components, we can obtain the transformation laws for components of any superfield. We note that there is only one set of generators Q_α, as it must be in the case of the $\mathcal{N} = 1$ supersymmetry.

The (super)covariant derivative D_α of any superfield must be consistent with the supersymmetry, i.e. under supersymmetry transformations (3.10) it also should vary as a superfield, $\delta D_\alpha \Phi = D_\alpha \delta \Phi$. To satisfy this condition, the D_α must anticommute with the supersymmetry generators Q_β, i.e. $\{D_\alpha, Q_\beta\} = 0$, and commute with the translation generators $P_{\alpha\beta} = i\partial_{\alpha\beta}$. Also, spinor supercovariant derivatives must be linear in the simple derivatives ∂_α, $\partial_{\alpha\beta}$ in order to satisfy the Leibnitz rule. All these properties are satisfied if the D_α looks like

$$D_\alpha = \partial_\alpha + i\theta^\beta \partial_{\beta\alpha}. \tag{3.11}$$

The spinor supercovariant derivatives D_α defined in such a way possess the following properties:

$$\{D_\alpha, D_\beta\} = 2i\partial_{\alpha\beta}; \quad [D_\alpha, D_\beta] = -2C_{\alpha\beta}D^2. \tag{3.12}$$

After summation of two these expressions we arrive at the following very important relation

$$D_\alpha D_\beta = i\partial_{\alpha\beta} - C_{\alpha\beta}D^2. \tag{3.13}$$

This identity is employed to carry out D-algebra transformations which are fundamental for simplifying the forms of the contributions of the supergraphs. Many examples will be presented further. One can also notice that the first expression in (3.12) can be interpreted in the sense that the superspace possesses a fundamental torsion since the anticommutator of the covariant derivatives is proportional to the supercovariant derivative as occurs in spacetimes with a torsion.

Let us obtain some other useful identities (some of details presented here can be found also in [37]; note, however, that our conventions differ from those ones used in [37]). First of all, let us note that the totally antisymmetric object with three spinor indices should vanish. Indeed, such an object is equal to zero if any two of its indices coincide, and since spinor indices can take only two values, that is, 1 and 2, this object should have at least two coinciding indices. If such an object is constructed through antisymmetrizing the product $D_\alpha D_\beta D_\gamma$, we get

$$D_\alpha D_\beta D_\gamma + D_\beta D_\gamma D_\alpha + D_\gamma D_\alpha D_\beta - D_\alpha D_\gamma D_\beta - D_\beta D_\alpha D_\gamma$$
$$- D_\gamma D_\beta D_\alpha = 0. \tag{3.14}$$

Contracting this expression to $C^{\beta\gamma}$ we get

$$2(D_\alpha D_\beta D^\beta + D_\beta D^\beta D_\alpha + D^\beta D_\alpha D_\beta) = 0. \tag{3.15}$$

Then, in the first term we substitute the identity $D_\alpha D_\beta = -D_\beta D_\alpha + \{D_\alpha, D_\beta\}$, and in the second one we substitute the identity $D^\beta D_\alpha = -D_\alpha D^\beta + \{D^\beta, D_\alpha\}$. Afterwards, the anticommutator terms cancel each other, and we rest with

$$D^\beta D_\alpha D_\beta = 0. \tag{3.16}$$

This is a very important identity which role is similar to the properties of the projecting operators in four-dimensional superfield supersymmetry which will be discussed in the next chapter. Applying this identity to (3.15) we find another important identity

$$\{D_\alpha, D^2\} = 0. \tag{3.17}$$

Using the (3.13) and the (3.17) we can derive one more important relation

$$(D^2)^2 = \Box. \tag{3.18}$$

These properties of the supercovariant derivatives can be used for constructing of the superfields.

One more very important property of the supercovariant derivatives and the delta function which can be checked by direct applying the spinor supercovariant derivatives on the delta function is

$$\delta_{12} D^2 \delta_{12} = \delta_{12}. \tag{3.19}$$

It is straightforward to verify as well that

$$\delta_{12} D_\alpha \delta_{12} = 0. \tag{3.20}$$

The most important superfields used in the known field theory models in the three-dimensional superspace are the scalar and the spinor ones. The scalar superfield is defined in the form of the following θ expansion:

$$\phi(x, \theta) = \varphi(x) + \theta^\alpha \psi_\alpha(x) - \theta^2 F(x). \tag{3.21}$$

Its components can be also defined as the projections:

$$\begin{aligned}
\varphi(x) &= \phi(x, \theta)|; \\
\psi_\alpha(x) &= D_\alpha \phi(x, \theta)|; \\
F(x) &= D^2 \phi(x, \theta)|.
\end{aligned} \tag{3.22}$$

Here and further the symbol | means that the Grassmannian coordinates θ in the corresponding expression are put to zero after taking derivatives. In the expression above, the $\varphi(x)$ is the usual scalar field, $\psi_\alpha(x)$ is the spinor one, and $F(x)$ is the auxiliary field, whose equations of motion in the usual theory of the scalar superfield without higher derivatives are just constraints, so, it can be completely eliminated from the theory on the mass shell. The supersymmetry transformations for the component fields can be obtained from the projections:

$$
\begin{aligned}
\delta\varphi(x) &= \delta\phi(x,\theta)| = -i\epsilon^{\alpha} Q_{\alpha}\phi(x,\theta)| = \epsilon^{\alpha} D_{\alpha}\phi(x,\theta)| = \epsilon^{\alpha}\psi_{\alpha}(x);\\
\delta\psi_{\alpha}(x) &= \delta D_{\alpha}\phi(x,\theta)| = -i\epsilon^{\beta} Q_{\beta} D_{\alpha}\phi(x,\theta)| = -\epsilon^{\beta} D_{\alpha} D_{\beta}\phi(x,\theta)| =\\
&= -\epsilon^{\beta}(i\partial_{\alpha\beta} - C_{\alpha\beta} D^2)\phi(x,\theta)| = -\epsilon^{\beta}(i\partial_{\alpha\beta}\varphi(x) - C_{\alpha\beta} F(x));\\
\delta F(x) &= \delta D^2\phi(x,\theta)| = -i\epsilon^{\alpha} Q_{\alpha} D^2\phi(x,\theta)| = \epsilon^{\alpha} D_{\alpha} D^2\phi(x,\theta)| =\\
&= -\frac{1}{2}\epsilon^{\alpha}\{D_{\alpha}, D_{\beta}\} D^{\beta}\phi(x,\theta)| = -i\epsilon^{\alpha}\partial_{\alpha\beta}\psi^{\beta}.
\end{aligned} \tag{3.23}
$$

We conclude that the spinor field $\psi_{\alpha}(x)$ related with the scalar one $\varphi(x)$ through a supersymmetry transformation, thus, we refer to the ψ_{α} as to the superpartner of φ. It is important to note also that the superfield (3.21) describes two bosonic degrees of freedom, those ones of $\varphi(x)$ and $F(x)$, and two fermionic ones, corresponding to two components of the spinor ψ_{α}, i.e. the numbers of bosonic and fermionic degrees of freedom are equal. This is a common rule for all superfields and all supersymmetric field theories. The set of all components of the (scalar) superfield, in this case— $(\varphi(x), \psi_{\alpha}(x), F(x))$, is called the (scalar) supermultiplet.

Another important superfield is the spinor one defined as

$$
A_{\alpha}(x,\theta) = \chi_{\alpha}(x) - \theta_{\alpha} B(x) + i\theta^{\beta} V_{\beta\alpha}(x) - 2\theta^2[\lambda_{\alpha}(x) + \frac{i}{2}\partial_{\alpha\beta}\chi^{\beta}(x)], \tag{3.24}
$$

therefore its components are

$$
\begin{aligned}
\chi_{\alpha}(x) &= A_{\alpha}(x,\theta)|;\\
B(x) &= \frac{1}{2} D^{\alpha} A_{\alpha}(x,\theta)|;\\
V_{\alpha\beta}(x) &= -\frac{i}{2} D_{(\alpha} A_{\beta)}(x,\theta)|;\\
\lambda_{\alpha} &= \frac{1}{2} D^{\beta} D_{\alpha} A_{\beta}(x,\theta)|.
\end{aligned} \tag{3.25}
$$

Here the $V_{\alpha\beta}(x)$ (we note again that it is symmetric) is a bispinor form of the usual vector field (in the most interesting case—the gauge one), $\lambda_{\alpha}(x)$ is its superpartner (photino), and $\chi_{\alpha}(x)$ and $B(x)$ are the auxiliary fields.

These scalar and spinor superfields are the basic ingredients for constructing the most popular field theory models in the three-dimensional superspace. In principle, other superfields (for example bispinor ones) can be also introduced.

3.2 Field Theory Models

Now, let us introduce the models involving the scalar and spinor superfields. In this section, we proceed mostly in a manner similar to [23].

1. The action for models involving only scalar fields has the simple form

$$S = \int d^5z [\frac{1}{2}\Phi D^2\Phi - \frac{m}{2}\Phi^2 + f(\Phi)] \tag{3.26}$$

where $f(\Phi)$ is an arbitrary function of the scalar superfield (by the reasons of renormalizability it must have no more than fourth order in Φ). The case of the complex scalar superfield does not essentially differ. Here and further we denote the superspace measure $d^5z \equiv d^3x d^2\theta$.

The component form of this action can be obtained in the following way: since the integration and the differentiation are equivalent, $\int d^2\theta f = D^2 f|$, after integrating by parts in the kinetic term one has

$$S = \int d^3x [\frac{1}{2}D^2(\Phi D^2\Phi) - \frac{m}{2}D^2(\Phi^2) + D^2 f(\Phi)]|, \tag{3.27}$$

that is,

$$S = \int d^3x [\frac{1}{2}(D^2\Phi D^2\Phi + \Phi\Box\Phi + \frac{1}{2}D^\alpha\Phi D_\alpha D^2\Phi) - \frac{m}{2}(2\Phi D^2\Phi + D^\alpha\Phi D_\alpha\Phi) +$$
$$+ (\frac{1}{2}f''(\Phi)D^\alpha\Phi D_\alpha\Phi + f'(\Phi)D^2\Phi)]|, \tag{3.28}$$

which, with use of (3.22), yields

$$S = \int d^3x [\frac{1}{2}F^2 - \frac{1}{2}i\psi_\alpha\partial^{\alpha\beta}\psi_\beta + \frac{1}{2}\varphi\Box\varphi -$$
$$- m(\psi^2 + \varphi F) + f''(\varphi)\psi^2 + f'(\varphi)F]. \tag{3.29}$$

This is the general approach for reduction of a superfield action to components. We see that the masses of all fields $\varphi(x)$, $\psi_\alpha(x)$, $F(x)$ composing this supermultiplet are equal. This is a common situation taking place in various supersymmetric field theories—in any supermultiplet, masses of all component fields are equal.

As we have already mentioned, the auxiliary field F can be eliminated with use of the equation of motion

$$F = (m\varphi - f'(\varphi)). \tag{3.30}$$

which gives

$$S = \int d^3x \left[\frac{1}{2} \varphi \Box \varphi - \frac{1}{2} (m\varphi - f'(\varphi))^2 - \frac{1}{2} i \psi_\alpha \partial^{\alpha\beta} \psi_\beta - (m - f''(\varphi)) \psi^2 \right].$$

(3.31)

This is the theory of the self-coupled scalar field φ interacting also with the spinor ψ. In other words, it is a supersymmetric extension of the scalar field theory. In particular, for $f(\Phi) = \lambda \Phi^4$ (this is the higher possible renormalizable self-coupling of the scalar superfield, we note that the mass dimension of Φ is $\frac{1}{2}$ as well as of the derivative D_α, and of d^5z is -2) we get the usual renormalizable coupling $\lambda \varphi^6$ as one of the terms of interaction which is present in the theory.

2. The construction of action for the spinor superfield is more involved. The reason is that the spinor multiplet turns out to describe supersymmetric three-dimensional gauge models, hence we must formulate the gauge invariant model for this superfield, since it contains the vector $V_{\alpha\beta}(x)$ as one of the components.

We start with introduction of the three-dimensional gauge transformations for the scalar superfield Φ (cf. [23]):

$$\Phi \to e^{iK} \Phi, \quad \bar{\Phi} \to \bar{\Phi} e^{-iK},$$

(3.32)

where K is a superfield gauge transformation parameter. For a constant K the kinetic term $\frac{1}{2} \int d^5z \, D^\alpha \bar{\Phi} D_\alpha \Phi$ is evidently invariant under these transformations. Then we introduce a (gauge) covariant derivative

$$\nabla_\alpha \Phi = (D_\alpha + i A_\alpha) \Phi,$$

(3.33)

which under the transformation (3.32) carried out together with the following transformations for the A_α superfield:

$$A_\alpha \to A_\alpha - D_\alpha K$$

(3.34)

is transformed as

$$\nabla_\alpha \Phi \to e^{iK} \nabla_\alpha \Phi.$$

(3.35)

The complex conjugate expression $(\nabla_\alpha \Phi)^*$ is, in a similar manner, transformed by the factor e^{-iK}. Therefore the expression

$$\nabla^\alpha \Phi (\nabla_\alpha \Phi)^*$$

(3.36)

is invariant under the transformations (3.32), (3.34). It is natural to consider it as a simplest Lagrangian for the scalar field coupled to the gauge one, introducing thus the following action:

$$S_m = \int d^5 z \left[-\frac{1}{2} (D^\alpha \Phi + i \Phi A^\alpha)(D_\alpha \bar{\Phi} - i A_\alpha \bar{\Phi}) - m \Phi \bar{\Phi} \right], \qquad (3.37)$$

Taking into account (3.32), (3.35) we can write down the following formal transformation law for the ∇_α:

$$\nabla_\alpha \rightarrow e^{iK} \nabla_\alpha e^{-iK}. \qquad (3.38)$$

The covariant derivatives ∇_α represent themselves as a base for constructing the superfield strengths.

We impose the following anticommutation relation which is a straightforward covariant generalization of first relation in (3.12):

$$\{\nabla_\alpha, \nabla_\beta\} = 2i \nabla_{\alpha\beta}. \qquad (3.39)$$

If we suggest for any covariant derivative ∇_A (with both $A = \alpha$ and $A = \alpha\beta$) the relation $\nabla_A = D_A + i\Gamma_A$, we get $\Gamma_\alpha = A_\alpha$ and $\Gamma_{\alpha\beta} = -\frac{i}{2} D_{(\alpha} A_{\beta)}$.

Then we suggest that the Bianchi identities on the ∇_A are valid:

$$[\nabla_{[A}, [\nabla_B, \nabla_{C]}\}\} = 0, \qquad (3.40)$$

where the anticommutator (and symmetrization over indices) is suggested between two fermionic objects whereas the commutator (and antisymmetrization over indices)—in all other cases. We also introduce the general curvature-torsion definition:

$$[\nabla_A, \nabla_B\} = T_{AB}^C \nabla_C - i F_{AB}, \qquad (3.41)$$

where T_{AB}^C is a torsion (note that unlike of the "common" flat space the superspace possesses the nontrivial intrinsic torsion even for "simple" covariant derivatives $D_\alpha, \partial_{\alpha\beta}$), and F_{AB} is a curvature. Suggesting in (3.40) the set $A, B, C = \alpha, \beta, \gamma$ we get

$$[\nabla_\alpha, \{\nabla_\beta, \nabla_\gamma\}] + [\nabla_\beta, \{\nabla_\gamma, \nabla_\alpha\}] + [\nabla_\gamma, \{\nabla_\alpha, \nabla_\beta\}] = 0, \qquad (3.42)$$

which implies

$$[\nabla_{(\alpha}, \nabla_{\beta\gamma)}] \equiv -i F_{(\alpha, \beta\gamma)} = 0. \qquad (3.43)$$

Also, it follows from (3.42) that $T_{\alpha,\beta\gamma}^D = 0$. Then, splitting the $F_{\alpha,\beta\gamma}$ into the irreducible representations we get

$$F_{\alpha,\beta\gamma} = \frac{1}{6} F_{(\alpha,\beta\gamma)} - \frac{1}{3} C_{\alpha(\beta|} F^\delta_{,\,\delta|\gamma)}, \qquad (3.44)$$

where the symbol | means that the index δ is not affected by the symmetrization. Since $F_{(\alpha,\beta\gamma)} = 0$, we get $F_{\alpha,\beta\gamma} = iC_{\alpha(\beta}W_{\gamma)}$ with $W_\gamma = \frac{i}{3}F^\delta_{,\delta\gamma}$ Applying the definition of F_{AB} through the covariant derivatives (3.41) and taking into account that $T^D_{\alpha,\beta\gamma} = 0$ which naturally follows from the Bianchi identities above, we arrive at the following expression for the W_α which henceforth will be called the (three-dimensional) superfield strength:

$$W_\alpha = \frac{1}{2}D^\beta D_\alpha A_\beta. \tag{3.45}$$

This object is invariant under the transformations (3.34) which with taking into account the component structure of A_β superfield (3.25) correspond to the following variation of the component fields:

$$\delta\chi_\alpha = -\sigma_\alpha, \quad \delta B = -\tau, \quad \delta V_{\alpha\beta} = -\partial_{\alpha\beta}\omega, \quad \delta\lambda_\alpha = 0, \tag{3.46}$$

where for the given superfield gauge parameter K its components are defined as $\omega = K|, \sigma_\alpha = D_\alpha K|, \tau = D^2 K|$. The remarkable fact is that the transformation (3.34) of the superfield A_α corresponds to the common gradient transformation for its vector component $V_{\alpha\beta}$, so, it is indeed a consistent superfield generalization of the gauge transformation. We also note that by an appropriate choice of the gauge parameter K (and hence of its components σ_α, τ) we can completely gauge away the components χ_α and B. Such a gauge choice providing $\chi_\alpha = B = 0$ is called the Wess-Zumino (WZ) gauge. Its advantage consists in vanishing of all terms involving third and higher powers of the A_α superfield itself in the vertices of interaction but it implies in breaking of supersymmetry, with only some residual supersymmetry persists in this case. It must be noted that when the WZ gauge is applied, the terms involving derivatives of A_α must be considered in a more careful manner, for example, while the non-Abelian term $\{A^\alpha, A^\beta\}\{A_\alpha, A_\beta\}$ vanishes in this gauge, the $\{A^\alpha, D^\gamma A^\beta\}\{A_\alpha, D_\gamma A_\beta\}$ does not vanish.

The component structure of the W_α strength looks as follows:

$$W_\alpha = \lambda_\alpha + \theta^\beta f_{\alpha\beta} + i\theta^2 \partial_{\alpha\beta}\lambda^\beta, \tag{3.47}$$

i.e. the W_α involves only the tensor $f_{\alpha\beta} = \frac{1}{2}(\sigma^{mn})_{\alpha\beta}F_{mn}$. Here $F_{mn} = \partial_m V_n - \partial_n V_m$ is the usual stress tensor, and $\sigma^{mn} = [\gamma^m, \gamma^n]$, and it should be noted that σ^{mn} with two upper or two lower spinor indices is a symmetric matrix with respect to the spinor indices. This follows from the fact that the Dirac matrices used within this section satisfy the relation: $\gamma^m\gamma^n = \eta^{mn} - \epsilon^{mnl}\gamma_l$). This stress tensor is evidently gauge invariant, and its component expansion is not modified even after imposing of the WZ gauge. The components of W_α can be defined as follows:

$$\lambda_\alpha = W_\alpha|;$$
$$f_{\alpha\beta} = D_\alpha W_\beta|. \tag{3.48}$$

We note the transversality of the W^α following from (3.45):

$$D^\alpha W_\alpha = 0, \tag{3.49}$$

which is a superfield analogue of the known property of the dual vector $F^m = \frac{1}{2}\epsilon^{mnl}F_{nl}$:

$$\partial_m F^m = 0. \tag{3.50}$$

The most natural, gauge invariant kinetic term for the action of the spinor superfield is hence

$$S_g = \frac{1}{2g^2}\int d^5z\, W^\alpha W_\alpha, \tag{3.51}$$

which, because of (3.47), can be easily shown to give in components

$$S_g = -\frac{1}{g^2}\int d^3x\left(\frac{1}{2}f^{\alpha\beta}f_{\alpha\beta} + \lambda^\alpha i\partial_{\alpha\beta}\lambda^\beta\right). \tag{3.52}$$

Coupling of the gauge superfield to matter is given by the term

$$S_m = -\frac{1}{2}\int d^5z\,\nabla^\alpha\Phi(\nabla_\alpha\Phi)^* = -\frac{1}{2}\int d^5z\,(D^\alpha + iA^\alpha)\Phi(D_\alpha - iA_\alpha)\Phi^*, \tag{3.53}$$

whose component content is

$$
\begin{aligned}
S_m = \int d^3x\Big[&(F\bar F - i\bar\psi_\alpha\partial^{\alpha\beta}\psi_\beta + \bar\varphi\Box\varphi + \\
&+ iV^{\alpha\beta}\varphi\,\overset{\leftrightarrow}{\partial}_{\alpha\beta}\bar\varphi - \frac{1}{2}\varphi V^{\alpha\beta}V_{\alpha\beta}\bar\varphi - V^{\alpha\beta}(\psi_\alpha\bar\psi_\beta + \bar\psi_\alpha\psi_\beta) + \lambda^\alpha(\varphi\bar\psi_\alpha + \bar\varphi\psi_\alpha) - \\
&- (\psi^\alpha\bar F - \bar\psi^\alpha F)\chi_\alpha - (\partial_{\alpha\beta}\psi^\alpha\bar\varphi - \partial_{\alpha\beta}\bar\psi^\alpha\varphi + \partial_{\alpha\beta}\bar\varphi\psi^\alpha + \partial_{\alpha\beta}\varphi\bar\psi^\alpha)\chi^\beta - \\
&- \frac{1}{2}(\varphi B^2\bar\varphi - \varphi\bar\varphi\chi^\alpha\lambda_\alpha + \chi^\alpha\chi_\alpha(F\bar\varphi + \varphi\bar F))\Big].
\end{aligned}
\tag{3.54}
$$

We note that only the terms in two first lines do not vanish in the WZ gauge. Presence of the additional terms, or, as is the same, additional interaction vertices, implies in the known difference between the results obtained within superfield approach and component approach—it must be noted that most of papers devoted to the component calculations use the "simplified" supersymmetric formulation of the theories which is effectively obtained by imposing of the WZ gauge. The reason is that such a theory really possesses a residual supersymmetry and contains much less terms. However, such a formulation cannot be obtained by direct projecting of the superfield action into components. This difference is a quite known phenomenon, e.g. in four-dimensional SYM theory, imposing of the WZ gauge allows to truncate a nonpolynomial expansion of the action. However, in this case only the residual supersymmetry survives, which makes the superfield description to be a bit senseless (for a discussion of noncovariant gauges in superfield theories see e.g. [39]).

One more gauge invariant action in the three-dimensional superspace is the superfield analogue of the Chern-Simons term:

$$S_{CS} = \frac{m}{2g^2} \int d^5z A^\alpha W_\alpha, \tag{3.55}$$

which component structure is

$$S_{CS} = \frac{m}{g^2} \int d^3x (V^{\alpha\beta} f_{\alpha\beta} - \lambda^\alpha \lambda_\alpha). \tag{3.56}$$

As we noticed already, the $f_{\alpha\beta}$ is a vector dual to the stress tensor.

3. There is also an alternative free action for the spinor superfield initially introduced in [40]. In this case, we have two spinor fields, Ψ^α and $\bar{\Psi}^\alpha$ whose component structure is

$$\Psi^\alpha = \psi^\alpha + \theta^\alpha b + i\theta_\beta b^{\beta\alpha} - \theta^2 \phi^\alpha;$$
$$\bar{\Psi}^\alpha = \bar{\psi}^\alpha + \theta^\alpha \bar{b} + i\theta_\beta \bar{b}^{\beta\alpha} - \theta^2 \bar{\phi}^\alpha. \tag{3.57}$$

We can introduce the Dirac-like action for these fields:

$$S = - \int d^5z \bar{\Psi}^\alpha (i\partial_{\alpha\beta} - MC_{\alpha\beta})\Psi^\beta. \tag{3.58}$$

The corresponding action for the component fields looks like

$$S_M = S_M^{(1/2)} + S_M^{(1)}, \tag{3.59}$$

where

$$S_M^{(1/2)} = \int d^3x \left[\bar{\phi} \left(i\gamma^m \partial_m - M \right) \psi + \bar{\psi} \left(i\gamma^m \partial_m - M \right) \phi \right], \tag{3.60}$$

$$S_M^{(1)} = - \int d^3x \left[\frac{1}{2}\varepsilon^{mnp}\bar{b}_m \partial_n b_p + \frac{M}{2} \bar{b}^m b_m + \bar{b}\partial^m b_m + b\partial^m \bar{b}_m - 2M\bar{b}b \right], \tag{3.61}$$

We can eliminate the auxiliary field b using its equation of motion, thus

$$S_M^{(1)} = - \int d^3x \left[\frac{1}{2}\varepsilon^{mnp}\bar{b}_m \partial_n b_p + \frac{M}{2} \bar{b}^m b_m - \frac{1}{2M} \left(\partial^m \bar{b}_m \right) \left(\partial^m b_m \right) \right], \tag{3.62}$$

This action can be treated as a gauge-fixed Chern-Simons theory, with a Proca mass term. However, the theory (3.58), up to now, has been discussed only within the duality context [40].

3.3 Non-Abelian Gauge Models

Here we generalize the formulation of the three-dimensional superfield gauge theories to the case of the non-Abelian gauge group. To develop it we suggest that the gauge superfield A^β takes values in a Lie algebra: $A^\beta = A^{\beta A} T^A$, where T^A are the generators of the corresponding Lie group. It should be emphasized that this superfield is fermionic. We suggest the scalar (matter) superfield also to be the Lie-algebra valued. Again, we follow the methodology of [23]. We start with the action of matter coupled to the gauge superfield in the non-Abelian case which can be introduced as

$$S_m = -\frac{1}{2} \int d^5 z (D^\alpha \Phi + i[\Phi, A^\alpha])(D_\alpha \bar{\Phi} - i[A_\alpha, \bar{\Phi}]), \qquad (3.63)$$

with the gauge transformations are

$$\Phi \to e^{-iK} \Phi e^{iK}, \quad \bar{\Phi} \to e^{-iK} \bar{\Phi} e^{iK}, \quad A_\alpha \to e^{-iK} A_\alpha e^{iK} - i e^{-iK} (D_\alpha e^{iK}). \qquad (3.64)$$

This is the case of the coupling of the matter to the gauge field in the adjoint representation where the matter is Lie-algebra valued as well as the gauge field. The transformation above for the A_α field in the infinitesimal form looks like

$$\delta A_\alpha = D_\alpha K + i[A_\alpha, K]. \qquad (3.65)$$

The introduction of the covariant derivatives is carried out as above:

$$\nabla^\alpha \Phi = D^\alpha \Phi + i[\Phi, A^\alpha]. \qquad (3.66)$$

Applying the identities (3.39), (3.42) as above, we again arrive at

$$W_\gamma = -\frac{i}{3} C^{\alpha\beta} F_{\alpha,\beta\gamma}, \qquad (3.67)$$

with the stress tensor F_{AB} is defined in (3.41). However, we must take into account that now the A^α superfields, as well as their derivatives, do not (anti)commute more, hence the $F_{\alpha,\beta\gamma} = i[\nabla_\alpha, \nabla_{\beta\gamma}]$ is nonlinear in A_α. Since it follows from (3.39) that

$$\Gamma_{\alpha\beta} = -\frac{i}{2} (D_{(\alpha} A_{\beta)} - i\{A_\alpha, A_\beta\}), \qquad (3.68)$$

we get

$$W_\alpha = \frac{1}{2} D^\beta D_\alpha A_\beta - \frac{i}{2}[A^\beta, D_\beta A_\alpha] - \frac{1}{6}[A^\beta, \{A_\beta, A_\alpha\}]. \tag{3.69}$$

The W_α defined in such a way is not invariant under transformation (3.64). Instead, it is transformed covariantly by the rule

$$W_\alpha \to e^{-iK} W_\alpha e^{iK}, \tag{3.70}$$

or, in the infinitesimal form,

$$W_\alpha \to W_\alpha + i[W_\alpha, K]. \tag{3.71}$$

Now, let us consider the Bianchi identity

$$\{\nabla_\alpha, [\nabla_\beta, \nabla_{\gamma\delta}]\} + \{\nabla_\beta, [\nabla_{\gamma\delta}, \nabla_\alpha]\} + [\nabla_{\gamma\delta}, \{\nabla_\alpha, \nabla_\beta\}] = 0. \tag{3.72}$$

After contractions with the symbols $C^{\alpha\gamma}$ and $C^{\beta\delta}$ it gives

$$\{\nabla^{(\alpha}, [\nabla^{\beta)}, \nabla_{\alpha\beta}]\} = -6\{\nabla^\alpha, W_\alpha\} = 0. \tag{3.73}$$

This is the non-Abelian generalization of the transversality condition (3.49).

The most natural definition of the non-Abelian gauge invariant action, i.e. the action of the SYM theory is similar to the Abelian one:

$$S_{SYM} = \frac{1}{2g^2} \mathrm{tr} \int d^5 z W^\alpha W_\alpha. \tag{3.74}$$

Using the expression (3.69) and adding the gauge-fixing action

$$S_{gf} = -\frac{1}{4g^2\xi} \mathrm{tr} \int d^5 z (D^\alpha A_\alpha) D^2 (D^\beta A_\beta), \tag{3.75}$$

one can write down the exact form of total action of A_α superfield:

$$S_{total} = \frac{1}{2g^2} \mathrm{tr} \int d^5 z \left[\frac{1}{2}(1 + \frac{1}{\xi}) A^\alpha \Box A_\alpha - \frac{1}{2}(1 - \frac{1}{\xi}) A^\alpha i \partial_{\alpha\beta} D^2 A^\beta \right] +$$
$$+ \frac{1}{g^2} \mathrm{tr} \int d^5 z \left[-\frac{i}{4} D^\gamma D^\alpha A_\gamma [A^\beta, D_\beta A_\alpha] - \frac{1}{12} D^\gamma D^\alpha A_\gamma [A^\beta, \{A_\beta, A_\alpha\}] - \right.$$
$$- \frac{1}{8}[A^\gamma, D_\gamma A^\alpha][A^\beta, D_\beta A_\alpha] + \frac{i}{12}[A^\gamma, D_\gamma A^\alpha][A^\beta, \{A_\beta, A_\alpha\}] +$$
$$\left. + \frac{1}{72}[A^\gamma, \{A_\gamma, A^\alpha\}][A^\beta, \{A_\beta, A_\alpha\}] \right]. \tag{3.76}$$

We already noted that the action (3.74) of the SYM theory formally replays the structure of the action for the supersymmetric Abelian gauge theory (3.51). For the Chern-Simons theory, however, the action is not a "direct" non-Abelian generalization of (3.55). Indeed, under the infinitesimal transformation (3.71), the "naive" non-Abelian generalization of the Chern-Simons action (3.55) $S_{naive} = \frac{m}{2g^2} \text{tr} \int d^5 z A^\alpha W_\alpha$, obtained by a direct promotion of the Abelian strength W_α to its non-Abelian analogue (3.69), acquires a variation

$$\delta S_{naive} = -\frac{m}{2g^2} \text{tr} \int d^5 z K D^\alpha W_\alpha, \tag{3.77}$$

but $D^\alpha W_\alpha \neq 0$ in the non-Abelian case, see (3.73). To cancel this variation we should add to the S_{naive} some new terms. As a result, the gauge invariant action takes the form

$$S_{CS} = \frac{m}{2g^2} \text{tr} \int d^5 z (A^\alpha W_\alpha + \frac{i}{6} \{A^\alpha, A^\beta\} D_\beta A_\alpha + \frac{1}{12} \{A^\alpha, A^\beta\} \{A_\alpha, A_\beta\}). \tag{3.78}$$

The ghost action in both these theories, that is, SYM and non-Abelian supersymmetric Chern-Simons theories, is the same since it can be obtained from the same gauge-fixing function $\chi = D^\alpha A_\alpha$ by use of the Faddeev-Popov prescription. It looks like

$$S_{gh} = \frac{1}{2g^2} \text{tr} \int d^5 z c' D^\alpha (D_\alpha c + i\{A_\alpha, c\}). \tag{3.79}$$

Here c, c' are the Faddeev-Popov ghosts, they are fermionic superfields as it must be. Their component structure is the same as of the usual scalar superfield (3.21). We note that the last term in the expression above must include an anticommutator to provide its vanishing for the Abelian gauge group.

Now we are in position to develop the perturbative approach for the superfield theories.

3.4 Quantum Description for the Superfield Models

Our aim here consists in the development of the perturbative approach for the theories described above, that is, scalar superfield model, SYM and super-Chern-Simons field theories.

We start with the introduction of the superfield propagators. As it is known from quantum field theory, the usual definition of the propagator in the theory with the generating functional $Z[J]$ is

$$G(z_1, z_2) = \left(\frac{1}{i} \frac{\delta}{\delta J(z_1)}\right) \left(\frac{1}{i} \frac{\delta}{\delta J(z_2)}\right) Z[J]|_{J=0}. \tag{3.80}$$

The equivalent definition of the propagator is the following one: if the operator characterizing the free theory (that is, the second functional derivative of the free action) is Δ, the propagator $G(z_1 - z_2)$ satisfies the equation

$$\Delta G(z_1 - z_2) = i\delta^5(z_1 - z_2). \tag{3.81}$$

In other words, the propagator in the theory characterized by the operator Δ is given by $G(z_1, z_2) = i\Delta^{-1}\delta^5(z_1 - z_2)$. Natural generalizations of this definition will be employed in the sequel for all superfield theory models we consider in this book.

For the scalar field theory (3.26) the propagator is

$$G(z_1, z_2) = \langle \Phi(z_1)\Phi(z_2)\rangle = i\frac{D^2 + m}{\Box - m^2}\delta(z_1 - z_2). \tag{3.82}$$

The ghost propagator is very similar:

$$G^{gh}(z_1, z_2) = \langle c(z_1)c'(z_2)\rangle = ig^2\frac{D^2}{\Box}\delta(z_1 - z_2). \tag{3.83}$$

We note that any ghost loop, despite the similarity of the ghost and scalar propagators, will carry an additional minus sign since ghosts are fermions.

For the QED, and similarly for the SYM theory where the only difference will consist in the presence of extra algebraic indices, because of the gauge invariance, we must fix the gauge by adding to the action (3.51) the gauge fixing term

$$S_{GF}^{QED} = -\frac{1}{4\xi g^2} \int d^5z (D^\alpha A_\alpha)D^2(D^\beta A_\beta), \tag{3.84}$$

which gives the propagator

$$G_{QED}^{\alpha\beta}(z_1, z_2) = \langle A^\alpha(z_1)A^\beta(z_2)\rangle = \frac{ig^2}{2\Box^2}[D^2 D^\beta D^\alpha - \xi D^2 D^\alpha D^\beta]\delta(z_1 - z_2). \tag{3.85}$$

After applying the identity (3.13), this propagator takes the form

$$G_{QED}^{\alpha\beta}(z_1, z_2) = \langle A^\alpha(z_1)A^\beta(z_2)\rangle =$$
$$= \frac{ig^2}{2}\Big[C^{\alpha\beta}\frac{1}{\Box}(\xi + 1) - \frac{1}{\Box^2}(\xi - 1)i\partial^{\alpha\beta}D^2\Big]\delta(z_1 - z_2). \tag{3.86}$$

The most important gauges are: $\xi = 1$, the Feynman gauge, where the propagator does not involve spinor derivatives; $\xi = -1$, where only second term of (3.86) survives, and $\xi = 0$, the Landau gauge which makes the propagator to be transversal, i.e. $D^\alpha G_{\alpha\beta}|_{\xi=0} = 0$. Nevertheless, sometimes other gauges are also useful, for example, it was shown in [41] that the supersymmetric three-dimensional scalar QED is finite in all loop orders at $\xi = -8$.

For the Chern-Simons theory, we also must add to the free action (3.55) the gauge fixing term which in this case we choose to be in the form

$$S_{GF}^{CS} = -\frac{m}{4\xi g^2} \int d^5 z (D^\alpha A_\alpha)(D^\beta A_\beta), \tag{3.87}$$

implying in the propagator

$$G_{CS}^{\alpha\beta}(z_1, z_2) = \frac{ig^2}{2m\Box}[D^\beta D^\alpha + \xi D^\alpha D^\beta]\delta(z_1 - z_2). \tag{3.88}$$

Its equivalent form is

$$G_{CS}^{\alpha\beta}(z_1, z_2) = \langle A^\alpha(z_1) A^\beta(z_2) \rangle = \frac{ig^2}{2m\Box}\Big[C^{\alpha\beta}D^2(1-\xi) + (1+\xi)i\partial^{\alpha\beta}\Big]\delta(z_1 - z_2). \tag{3.89}$$

Some authors, instead of the gauge-fixing term (3.87), suggest to add another gauge-fixing term

$$S_{GF2}^{CS} = -\frac{m}{4\xi g^2} \int d^5 z (D^\alpha A_\alpha) D^2 (D^\beta A_\beta), \tag{3.90}$$

with ξ is now a parameter with a non-zero mass dimension. However, this choice, corresponding to the propagator

$$G_{CS2}^{\alpha\beta}(z_1, z_2) = \frac{ig^2}{2m\Box}[D^\beta D^\alpha + \xi \frac{D^2}{\Box} D^\alpha D^\beta]\delta(z_1 - z_2). \tag{3.91}$$

with the same ξ-independent part as that one of (3.88), does not imply an essentially different situation.

Within the three-dimensional Feynman diagrams we consider in this chapter, the scalar propagators and legs will be represented by solid lines, the gauge propagators and legs—by wavy lines, and the ghost propagators—by dashed lines.

It is instructive to give here the inverse operator for the generic one looking like $\Delta^{\alpha\beta} = A_1 D^\alpha D^\beta + A_2 D^\beta D^\alpha$. We suggest that both A_1 and A_2 commute with the product $D_\alpha D_\beta$, being either constants or functions of D^2 and space-time derivatives. Supposing that the inverse operator $Q_{\alpha\beta}$ by the definition satisfies the relation $\Delta^{\alpha\beta} Q_{\beta\gamma} = \delta_\gamma^\alpha$, we will expect it to have the form:

$$Q_{\beta\gamma} = X_1 D_\beta D_\gamma + X_2 D_\gamma D_\beta. \tag{3.92}$$

We carry out the straightforward multiplication of Δ and Q. The property $D^\alpha D_\beta D_\alpha = 0$ cancels two of four terms in this product. We simplify remaining ones with the use of the key identity (3.13), compare the factors accompanying δ_γ^α and ∂_γ^α in both sides of the equation and arrive at

$$2\Box(A_1 X_1 + A_2 X_2) = -1;$$
$$A_1 X_1 - A_2 X_2 = 0. \tag{3.93}$$

From this system, we find the coefficients X_1 and X_2, writing down the operator $Q_{\beta\gamma}$ as

$$Q_{\beta\gamma} = -\frac{1}{4\Box}\Big(\frac{1}{A_1} D_\beta D_\gamma + \frac{1}{A_2} D_\gamma D_\beta\Big). \tag{3.94}$$

It is easy to see that the cases of the superfield QED and the superfield Chern-Simons theory given above match this expression.

The quantum contributions as usual are described by the supergraphs arising from the standard generating functional:

$$Z[J] = \int D\Phi e^{i(\Phi\frac{\Delta}{2}\Phi + V(\Phi) + \Phi J)} \equiv e^{iV(\frac{1}{i}\frac{\delta}{\delta J})} e^{-\frac{i}{2}J\Delta^{-1}J} \tag{3.95}$$

To describe the general divergence structure in any superfield theory we can define the superficial degree of divergences (SDD). As usual, we assume that the common space-time derivative contributes 1 to the SDD. Therefore, because of the (3.12), each spinor supercovariant derivative, either in a propagator on in a vertex, corresponds to $\frac{1}{2}$. Then, the propagator of a scalar superfield yields -1 as well as the propagator of the Chern-Simons superfield, and that one of the gauge superfield in QED (and similarly— Maxwell-Chern-Simons and SYM theories) corresponds to -2. Each loop contributes 2 since any integration over d^3k yields 3, but a number of the D-factors which can be converted to momenta by the rule (3.13) is decreased by 2 in any loop because of the shrinking any loop to a point in the θ-space through the identity (3.19). We denote the number of vertices involving j gauge superfields A^α and no other superfields, as $V_A^{(j)}$, the number of vertices involving ghosts as V_c, the numbers of vertices involving scalar superfields with one and none spinor supercovariant derivatives as V_ϕ^D (which contributes $\frac{1}{2}$) and V_ϕ^0 respectively—here we suggest that the matter is coupled only to the gauge superfield just in the form given in (3.63). It follows from (3.76) that $V_A^{(3)}$ vertex involves three derivatives, the $V_A^{(4)}$—two derivatives, etc. Taking all together and using the topological identity $L + V - P = 1$, we can find that the SDD in the SYM theory with a scalar matter looks like (cf. [15]):

$$\omega = 2 - 2V_A^{(6)} - \frac{3}{2}V_A^{(5)} - V_A^{(4)} - \frac{1}{2}(V_A^{(3)} + V_c) - \frac{1}{2}E_\phi - \frac{1}{2}V_\phi^D - V_\phi^0, \tag{3.96}$$

whereas in the super-Chern-Simons theory—like

$$\omega = 2 - \frac{1}{2}(E_A + E_\phi). \tag{3.97}$$

Fig. 3.1 Contributions to
the two-point function of the
gauge superfield from the
matter sector

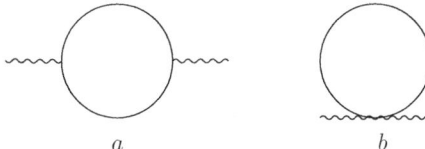

a b

Here the E_A and E_ϕ are the numbers of the external gauge and scalar legs. Therefore we find that the SYM theory is super-renormalizable (and finite beyond two
loops) whereas the super-Chern-Simons theory is renormalizable. Moreover, in the
framework of the dimensional regularization both theories are one-loop finite, since
the integral $\int \frac{d^3k}{k^2+m^2}$ is finite within this approach being proportional to $\Gamma(-1/2) =
-2\sqrt{\pi}$, and the one-loop logarithmic divergences in the odd-dimensional theories
vanish by symmetry reasons, except of the cases of very specific effective theories
where the propagators proportional to $\frac{1}{\sqrt{k^2}}$ are possible, the typical examples of such
theories are the nonlinear sigma model [42] and the CP^{N-1} model [15].

The contributions from the supergraphs are evaluated with help of the D-algebra
transformations whose aim consists in the reduction of the contribution to the supergraph to the single integral over $d^2\theta$ (the possibility of this reduction can be treated
as a some kind of the "nonrenormalization theorem" originally introduced in the
four-dimensional case [43]). These transformations are based on Leibnitz rule and
use identities (3.19), (3.20) and the similar ones. We note that for tadpole supergraphs we cannot put $D^2\delta(0) = 0$, really the expression of this type is treated as
$D^2\delta(0) \equiv D^2\delta_{12}|_{\theta_1=\theta_2} = 1$ (again, we note that, $\delta_{12} \equiv \delta(\theta_1 - \theta_2)$), this relation is
similar to (3.19). We also use the following relation allowing to transfer a derivative
from one argument of the delta function to another:

$$D_\alpha(\theta_1, k)\delta_{12} = -D_\alpha(\theta_2, -k)\delta_{12}. \qquad (3.98)$$

Let us give the typical example of the D-algebra transformations [15]. The matter
contribution to the two-point function spinor field A_α arising in the model (3.37) is
formed by two diagrams shown in Fig. 3.1.

The first graph, depicted in Fig. 3.1a, gives the following contribution:

$$i S_{1a}(p) = -\frac{1}{4} \int d^2\theta_1 d^2\theta_2 \int \frac{d^3k}{(2\pi)^3} A^\alpha(-p, \theta_1)A^\beta(p, \theta_2) \qquad (3.99)$$

$$\times \Big[D_{\alpha 1} \langle \phi(-k, \theta_1)\bar\phi(k, \theta_2)\rangle (\langle \bar\phi(k + p, \theta_1)\phi(-k - p, \theta_2)\rangle \overleftarrow{D}_{\beta 2})$$

$$- (D_{\alpha 1} \langle \phi(-k, \theta_1)\bar\phi(k, \theta_2)\rangle \overleftarrow{D}_{\beta 2})\langle \bar\phi(k + p, \theta_1)\phi(-k - p, \theta_2)\rangle \Big],$$

where the notation $D_{\gamma i}$ was used to indicate that the supercovariant derivative D_γ
is applied to the field whose Grassmannian argument is θ_i. Taking into account the
explicit form of the propagators (3.82), after Fourier transform we have

$$i S_{1a}(p) = \frac{1}{4} \int d^2\theta_1 d^2\theta_2 \int \frac{d^3k}{(2\pi)^3} A^\alpha(-p, \theta_1) A^\beta(p, \theta_2) \times$$

$$\times \left[\frac{D_{\alpha 1}(D_1^2 + m)}{k^2 + m^2} \delta_{12} \frac{(D_1^2 + m) D_{\beta 2}}{(k + p)^2 + m^2} \delta_{12} \right.$$

$$\left. - \frac{D_{\alpha 1}(D_1^2 + m) D_{\beta 2}}{k^2 + m^2} \delta_{12} \frac{D_1^2 + m}{(k + p)^2 + m^2} \delta_{12} \right]. \tag{3.100}$$

Integrating by parts some of the spinor derivatives and using the identity $D_{\beta 2}(k, \theta_2)$ $\delta_{12} = -D_{\beta 1}(-k, \theta_1)\delta_{12}$ (further, we will omit momentum arguments of spinor supercovariant derivatives), we arrive at

$$i S_{1a}(p) = \frac{1}{4} \int d^2\theta_1 d^2\theta_2 \int \frac{d^3k}{(2\pi)^3} I(k, p)$$

$$\times \left[2(D_1^2 + m)\delta_{12} D_{\alpha 1}(D_1^2 + m) D_{\beta 1}\delta_{12} A^\alpha(-p, \theta_1) A^\beta(p, \theta_2) \right.$$

$$\left. + (D_1^2 + m)\delta_{12}(D_1^2 + m) D_{\beta 1}\delta_{12}(D^\alpha A_\alpha)(-p, \theta_1) A^\beta(p, \theta_2) \right]. \tag{3.101}$$

where

$$I(k, p) = \frac{1}{(k^2 + m^2)[(k + p)^2 + m^2]}. \tag{3.102}$$

It is convenient to separate S_{1a} into two parts, $S_{1a} = S_{1a}^{(1)} + S_{1a}^{(2)}$, where $i S_{1a}^{(1)}$ and $i S_{1a}^{(2)}$ are associated to two terms in the large brackets of (3.101). Let us consider first $i S_{1a}^{(1)}$, which, after transporting D^2 from one of the propagators to other factors, becomes

$$i S_{1a}^{(1)}(p) = \frac{1}{4} \int d^2\theta_1 d^2\theta_2 \int \frac{d^3k}{(2\pi)^3} I(k, p)$$

$$\times \left[2m\delta_{12} D_{\alpha 1}(D_1^2 + m) D_{\beta 1}\delta_{12} A^\alpha(-p, \theta_1) A^\beta(p, \theta_2) \right.$$

$$\left. + 2\delta_{12} D_1^2 \Big(D_{\alpha 1}(D_1^2 + m) D_{\beta 1}\delta_{12} A^\alpha(-p, \theta_1) \Big) A^\beta(p, \theta_2) \right]. \tag{3.103}$$

Now we employ the identity $\{D_{\alpha 1}, D_1^2\} = 0$ which leads to

$$i S_{1a}^{(1)}(p) = \frac{1}{4} \int d^2\theta_1 d^2\theta_2 \int \frac{d^3k}{(2\pi)^3} I(k, p) \tag{3.104}$$

$$\times \left[2\delta_{12}(k^2 + m^2) D_{\alpha 1} D_{\beta 1}\delta_{12} A^\alpha(-p, \theta_1) A^\beta(p, \theta_2) \right.$$

$$\left. + 2\delta_{12}(-D_1^2 + m) D_{\alpha 1} D_{\beta 1}\delta_{12}(D^2 A^\alpha(-p, \theta_1)) A^\beta(p, \theta_2) \right].$$

The use of the relationship (3.13) now provides

$$
i S_{1a}^{(1)}(p) = \frac{1}{2} \int d^2\theta_1 d^2\theta_2 \int \frac{d^3k}{(2\pi)^3} I(k, p)
$$
$$
\times \Big[\delta_{12}(k^2 + m^2)(k_{\alpha\beta} - C_{\alpha\beta} D^2)\delta_{12} A^\alpha(-p, \theta_1) A^\beta(p, \theta_2)
$$
$$
+ \delta_{12}(-D^2 + m)(k_{\alpha\beta} - C_{\alpha\beta} D^2)\delta_{12}(D^2 A^\alpha(-p, \theta_1)) A^\beta(p, \theta_2) \Big]. \quad (3.105)
$$

The only terms giving non-zero contributions are those containing just one D^2 since $\delta_{12} D^2 \delta_{12} = \delta_{12}$, see (3.19). Indeed, by employing this identity and after integrating over θ_2 with the help of the delta function, we obtain

$$
i S_{1a}^{(1)}(p) = -\frac{1}{2} \int d^2\theta \int \frac{d^3k}{(2\pi)^3} I(k, p) \quad (3.106)
$$
$$
\times \Big[(k^2 + m^2) C_{\alpha\beta} A^\alpha(-p, \theta) A^\beta(p, \theta)
$$
$$
+ (k_{\alpha\beta} + m C_{\alpha\beta})(D^2 A^\alpha(-p, \theta)) A^\beta(p, \theta) \Big].
$$

The second term of (3.101) is

$$
i S_{1a}^{(2)}(p) = \frac{1}{4} \int d^2\theta_1 d^2\theta_2 \int \frac{d^3k}{(2\pi)^3} I(k, p)
$$
$$
\times \Big[(D_1^2 + m)\delta_{12}(D_1^2 + m) D_{\beta 1}\delta_{12}(D^\alpha A_\alpha)(-p, \theta_1) A^\beta(p, \theta_2) \Big]. \quad (3.107)
$$

In this expression we must keep only the term proportional to $D_1^2 \delta_{12}(D_1^2 + m) D_{\beta 1}\delta_{12}$ (the remaining part is a trace of an odd number of derivatives which clearly vanishes). Thus, after manipulations similar to those performed for $S_{1a}^{(1)}$, we find

$$
i S_{1a}^{(2)}(p) = -\frac{1}{4} \int d^2\theta \int \frac{d^3k}{(2\pi)^3} I(k, p) \Big[D^\gamma D^\alpha A_\alpha(-p, \theta)(k_{\gamma\beta} + m C_{\gamma\beta}) A^\beta(p, \theta) \Big].
$$
$$
(3.108)
$$

By adding (3.106) and (3.108) we can write the total contribution from Fig. 3.1a as

$$
i S_{1a}(p) = -\frac{1}{2} \int d^2\theta \int \frac{d^3k}{(2\pi)^3} I(k, p)
$$
$$
\times \Big[(k^2 + m^2) C_{\alpha\beta} A^\alpha(-p, \theta) A^\beta(p, \theta) + (k_{\alpha\beta} + m C_{\alpha\beta})(D^2 A^\alpha(-p, \theta)) A^\beta(p, \theta)
$$
$$
+ \frac{1}{2} D^\gamma D^\alpha A_\alpha(-p, \theta)(k_{\gamma\beta} + m C_{\gamma\beta}) A^\beta(p, \theta) \Big]. \quad (3.109)
$$

The algebraic manipulations for the graph Fig. 3.1b are much more simpler and yield

$$
i S_{1b}(p) = \frac{1}{2} \int \frac{d^3k}{(2\pi)^3} \frac{1}{(k + p)^2 + m^2} C_{\alpha\beta} A^\alpha(-p, \theta) A^\beta(p, \theta). \quad (3.110)
$$

The complete two-point vertex function for the A_α field is the sum of (3.109) and (3.110) and therefore reads

$$
i S_1(p) = -\frac{1}{2} \int d^2\theta \int \frac{d^3k}{(2\pi)^3} \frac{1}{(k^2 + m^2)[(k + p)^2 + m^2]} (k_{\gamma\beta} + m C_{\gamma\beta})
$$
$$
\times \left[(D^2 A^\gamma(-p, \theta)) A^\beta(p, \theta) + \frac{1}{2} D^\gamma D^\alpha A_\alpha(-p, \theta) A^\beta(p, \theta) \right]. \text{(3.111)}
$$

One can observe that the linear divergences presented in S_{1a} and S_{1b} are cancelled in the above result, and the logarithmic ones vanish by symmetry reasons, therefore this contribution is finite. However, this finiteness is based on the gauge symmetry rather than on the supersymmetry. Indeed, after integration over the internal momentum we find that this expression is proportional to the gauge invariant expression

$$
i S_1(p) = \int d^2\theta f(p)(W^\alpha(-p) W_\alpha(p) + 2m W^\alpha(p) A_\alpha(p)), \qquad \text{(3.112)}
$$

where

$$
f(p) = \int \frac{d^3k}{(2\pi)^3} \frac{1}{(k^2 + m^2)[(k + p)^2 + m^2]}.
$$

This expression, at $p \to 0$ (notice that $f(p)|_{p\to0} = \frac{1}{8\pi|m|}$), reproduces the expression for the quadratic Maxwell-Chern-Simons action. As a result, even if we consider the spinor A^α superfield as a purely external one, it acquires a nontrivial dynamics due to the one-loop correction. This is a key effect which also occurs in the supersymmetric CP^{N-1} model explicitly studied in [44] in the commutative case, and in [15] in the noncommutative one. We note that the only difference taking place in the noncommutative case consists in the modification of the factor $f(p)$, see (3.158).

3.5 Effective Action of the Three-Dimensional Superfield Theories and the Proper-Time Method

In this section we develop a prescription for calculating the superfield effective action within the three-dimensional superfield formalism. Here we adopt the formalism developed in Chap. 2, for superfield theories, and follow the approach proposed in [45].

Our starting point is the three-dimensional superfield theory described by the action (see e.g. [23]):

$$
S[\Phi] = \int d^5z \left[\frac{1}{2} \Phi D^2 \Phi - V(\Phi) \right], \qquad \text{(3.113)}
$$

where Φ is a scalar superfield. We will use again the loop expansion methodology. To do it, we make a shift in the field Φ:

$$\Phi \rightarrow \Phi + \sqrt{\hbar}\,\phi, \tag{3.114}$$

where, Φ is now a background (super)field, and ϕ is a quantum one. As a result, the classical action (3.113) takes the form

$$S[\Phi, \phi] = S[\Phi] + \int d^5 z \left(\hbar \frac{1}{2}\phi[D^2 - V''(\Phi)]\phi \right.$$
$$\left. - \hbar^{3/2} \frac{1}{3!} V'''(\Phi)\phi^3 - \hbar^2 \frac{1}{4!} V^{(IV)}(\Phi)\phi^4 \right)$$
$$+ \cdots, \tag{3.115}$$

where dots are for higher orders in ϕ which are irrelevant in the two-loop approximation. Following the definitions given in the Chap. 2, we see that the corresponding effective action $\Gamma[\Phi]$ is defined by the expression

$$\exp\left(\frac{i}{\hbar}\Gamma[\Phi]\right) = \int D\phi \, \exp\left(\frac{i}{\hbar} S[\Phi, \phi]\right). \tag{3.116}$$

The general structure of the effective action can be cast in a form of the derivative expansion:

$$\Gamma[\Phi] = \int d^5 z \, K(\Phi) + \int d^5 z \, F(D_\alpha \Phi, D^2 \Phi; \Phi) + \cdots, \tag{3.117}$$

where the $K(\Phi)$ is the term which we call the Kählerian effective potential by analogy with four-dimensional studies (see the next chapter), depending only on the superfield Φ but not on its derivatives, and the F is called auxiliary fields effective potential whose key property is its vanishing in the case when all derivatives of the superfields are equal to zero (these definitions have been firstly introduced in [27] for the four-dimensional superfield theories), and dots are for terms involving space-time derivatives of superfields. It is easy to see that F is at least of the second order in the auxiliary field of the scalar supermultiplet. It can be explicitly written as

$$F(D_\alpha \Phi, D^2 \Phi; \Phi) = F_2(\Phi)D^\alpha \Phi D_\alpha \Phi + \cdots, \tag{3.118}$$

where the $F_2(\Phi)$ is a function of Φ only but not of its derivatives, and the dots correspond to terms with four or more supercovariant derivatives. It is clear that this definition does not require to impose the condition $D_\alpha \Phi = 0$ which naively can be treated as a supercovariant analogue of the usual requirement for a background superfield to be constant, but known to imply in strong conceptual difficulties (see e.g. [46]). It is sufficient to use only the standard condition for this superfield to be slowly varying in the space-time, $\partial_m \Phi = 0$.

We will work with a loop expansion for the effective action Γ,

$$\Gamma[\Phi] = S[\Phi] + \hbar \Gamma^{(1)}[\Phi] + \hbar^2 \Gamma^{(2)}[\Phi] + \ldots, \qquad (3.119)$$

for the Kählerian potential K,

$$K(\Phi) = -V(\Phi) + \sum_{L=1}^{\infty} \hbar^L K_L(\Phi), \qquad (3.120)$$

and similarly for F.

We start by considering the one-loop effective action in the form

$$\Gamma^{(1)} = \frac{i}{2} \text{Tr} \ln[D^2 - V''(\Phi)]. \qquad (3.121)$$

As a first approximation, let us restrict ourselves the Kählerian part of the effective action. From a formal viewpoint this restriction corresponds to disregarding all terms depending on derivatives of Φ, both space-time and spinor ones, and allows us to calculate the quantum corrections to $V(\Phi)$. In this case, we add a field independent term $\frac{i}{2} \text{Tr} \ln(D^2)$ and write

$$\Gamma^{(1)} = \frac{i}{2} \text{Tr} \ln[\Box - V''(\Phi) D^2]. \qquad (3.122)$$

This expression can be represented via the Schwinger proper-time representation [47, 48]:

$$\begin{aligned} \Gamma^{(1)} &= \frac{i}{2} \text{Tr} \int_0^\infty \frac{ds}{s} e^{is[\Box - V''(\Phi) D^2]} \\ &= \frac{i}{2} \int d^5 z \int_0^\infty \frac{ds}{s} e^{is[\Box - V''(\Phi) D^2]} \delta^5(z - z')|_{z=z'} . \end{aligned} \qquad (3.123)$$

Again, since we are calculating only the Kählerian part of the effective action, we have

$$\Gamma^{(1)} = \frac{i}{2} \text{Tr} \int d^5 z \int_0^\infty \frac{ds}{s} e^{-is V''(\Phi) D^2} e^{is\Box} \delta^5(z - z')|_{z=z'} , \qquad (3.124)$$

or, using that $(D^2)^2 = \Box$,

$$e^{-is V''(\Phi) D^2} = \sum_{n=0}^{\infty} \frac{[-is V''(\Phi)]^{2n+1}}{(2n+1)!} \Box^n D^2 + \ldots . \qquad (3.125)$$

Here the dots stand for terms which do not contribute to the integral. At this point, we can clearly emphasize the difference between the calculation of $\Gamma^{(1)}$ in three- and four-dimensional space-times. In four dimensions [8, 33], the $\Gamma^{(1)}$ is given by an expression similar to Eq. (3.124), but there are more independent structures involving supercovariant derivatives and chiral and antichiral background superfields. The calculation of the exponential similar to Eq. (3.125) involves the solving of a coupled set of differential equations, whose solutions can be found but are of a rather cumbersome form. In three dimensions the number of independent structures is much smaller, actually only terms involving a D^2 will be relevant to the calculation of the Kählerian contribution to the effective action. We will shortly show that these terms can be directly summed, thus providing a closed-form expression for $\Gamma^{(1)}$.

Let us now consider a function $U(x, x'; s) = e^{is\Box}\delta^3(x - x')$. Its key property is that

$$i\frac{\partial U}{\partial s} = -\Box U, \tag{3.126}$$

which allows us to obtain

$$\Box^n U(x, x'; s)|_{x=x'} \equiv \Box^n e^{is\Box}\delta^3(x - x')|_{x=x'} = \frac{\sqrt{i}}{8\pi^{3/2}}\left(-i\frac{d}{ds}\right)^n \frac{1}{s^{3/2}} =$$
$$= \frac{i^{n+1/2}}{8\pi^{3/2}}\frac{(2n+1)!!}{2^n s^{3/2+n}}. \tag{3.127}$$

From Eq. (3.125), after calculating the trace using that $D^2\delta^2(\theta - \theta')|_{z=z'} = 1$ and $\frac{(2n+1)!!}{(2n+1)!} = \frac{1}{(2n)!!} = \frac{1}{2^n n!}$, we obtain

$$\Gamma^{(1)} = \frac{i}{16\pi^{3/2}}\int d^5z \int_0^\infty \frac{ds}{s}\sum_{n=0}^\infty \frac{\left[-i\sqrt{i}\,V''(\Phi)\right]^{2n+1}}{4^n n!}s^{n-1/2}. \tag{3.128}$$

By performing the summation, we end up with

$$\Gamma^{(1)} = -\frac{i\sqrt{i}}{16\pi^{3/2}}\int d^5z\,V''(\Phi)\int_0^\infty \frac{ds}{s^{3/2}}e^{-is[\frac{(V''(\Phi))^2}{4}]}. \tag{3.129}$$

After an appropriate analytic continuation, we find that (3.129) is proportional to a gamma function, and finally arrive at

$$\Gamma^{(1)} = \frac{1}{16\pi}\int d^5z\left[V''(\Phi)\right]^2. \tag{3.130}$$

This is our final expression for the one-loop Kählerian effective action. We note its finiteness, and we can also observe that this expression holds also in the noncommutative case. Evidently, since the background superfield in our case is constant in the space-time, the Moyal product of these superfields reduces to the usual one (in higher loops, this is not so—the noncommutative deformation of the two-loop effective potential is discussed in [45]). It is interesting to note that this result, up to a constant multiplier, can be obtained without calculations. Indeed, it follows already from (3.121) that the one-loop effective potential is a function of $V''(\Phi)$ only. Since our result is naturally finite (as we already noted, in three-dimensional theories the one-loop results diverge only for theories with exotic effective dynamics, e.g. with propagators $\propto \frac{1}{\sqrt{k^2}}$), it cannot depend on any arbitrary parameter like the cutoff scale μ. Therefore, by dimensional reasons, the only form of the effective potential is $a \left[V''(\Phi) \right]^2$, where a is a some number.

We note that in principle one could elaborate the expression (3.121), for the constant background, by a much more straightforward way. Indeed, if one denotes $-V'' = \Psi$, the one-loop effective action becomes $\Gamma^{(1)} = \frac{i}{2} \mathrm{Tr} \ln[D^2 + \Psi]$, which is a function of Ψ. Then, one can consider

$$\frac{d\Gamma^{(1)}}{d\Psi} = \frac{i}{2} \mathrm{Tr} \frac{1}{D^2 + \Psi}, \tag{3.131}$$

which, after finding the inverse operator together with the Fourier transform, becomes

$$\frac{d\Gamma^{(1)}}{d\Psi} = -\frac{i}{2} \int d^5 z \int \frac{d^3 k}{(2\pi)^3} \frac{D^2 - \Psi}{k^2 + \Psi^2} \delta^5(z - z')|_{z=z'}. \tag{3.132}$$

The D-algebra is trivial, and after Wick rotation we have

$$\frac{d\Gamma^{(1)}}{d\Psi} = -\frac{1}{2} \int d^5 z \int \frac{d^3 k}{(2\pi)^3} \frac{1}{k^2 + \Psi^2} = \int d^5 z \frac{\Psi}{8\pi}, \tag{3.133}$$

Integrating this equation with respect to Ψ, we arrive at the expression (3.130) obtained above. Nevertheless, we remind that the proper-time method we formulated can be applied for a wide class of superfield theory models, and its significance is not exhausted by a scalar superfield theory considered here. Further in this section we discuss the case of contributions depending on derivatives of background superfields.

Now, let us go to two loops. Applying the expansion given in Eqs. (3.115)–(3.119), we can find that the expression for the two-loop effective action $\Gamma^{(2)}$ looks like,

$$\Gamma^{(2)} = -i \int D\phi \exp\left(\frac{i}{2}\phi[D^2 - V''(\Phi)]\phi \right) \left[\frac{1}{2} \left(\frac{1}{3!} V'''(\Phi)\phi^3 \right)^2 - \frac{1}{4!} V^{(IV)}(\Phi)\phi^4 \right]. \tag{3.134}$$

The two-loop contributions are schematically represented by two supergraphs given by Fig. 3.2.

Fig. 3.2 General structure
of two-loop contributions

(a)　　　　　(b)

Since we are interested in calculating the two-loop Kählerian contribution to the effective action, we again assume that $D^\alpha \Phi = 0$, so that the background field dependent mass $\Psi = -V''(\Phi)$ is constant, thus the simple propagator

$$\langle \phi(z_1)\phi(z_2)\rangle = -i\frac{D^2 - \Psi}{\Box - \Psi^2}\delta^5(z_1 - z_2),\tag{3.135}$$

can be used.

Thus, the contributions from diagrams depicted at Fig. 3.2a and b respectively, after trivial D-algebra transformations, look like

$$\Gamma_a^{(2)} = -\frac{1}{4}\int d^5z\,\left[V'''(\Phi)\right]^2\Psi\int\frac{d^3kd^3l}{(2\pi)^6}\frac{1}{(k^2+\Psi^2)(l^2+\Psi^2)[(k+l)^2+\Psi^2]},\tag{3.136}$$

and

$$\Gamma_b^{(2)} = -\frac{1}{4}\int d^5z\,V^{(IV)}(\Phi)\int\frac{d^3kd^3l}{(2\pi)^6}\frac{1}{(k^2+\Psi^2)(l^2+\Psi^2)}.\tag{3.137}$$

After integration, these expressions look like

$$\Gamma_a^{(2)} = \frac{1}{128\pi^2}\int d^5z\,\left[V'''(\Phi)\right]^2\Psi\left[\frac{1}{\epsilon}+\ln\frac{\Psi^2}{\mu^2}\right].\tag{3.138}$$

and

$$\Gamma_b^{(2)} = -\frac{1}{32\pi^2}\int d^5z\,V^{(IV)}(\Phi)\Psi^2.\tag{3.139}$$

We can add the corresponding two-loop counterterm to cancel the divergence.

Up to now, we have considered only the Kählerian part of the effective action. Let us describe the general procedure to obtain the one-loop effective potential taking into account the supercovariant derivatives of the background superfield. As we have already noticed, the one-loop effective action (3.122), up to the irrelevant additive constant, reads

$$\Gamma^{(1)} = \frac{i}{2}\text{Tr}\ln(D^2+\Psi).\tag{3.140}$$

Using the Schwinger representation, we can write this effective action as

$$\Gamma^{(1)} = \frac{i}{2} \int d^5 z \int \frac{ds}{s} e^{is(D^2+\Psi)} \delta^5(z-z')|_{z=z'}.$$

(3.141)

We then introduce the operator

$$\Omega(s) = e^{is(D^2+\Psi)},$$

(3.142)

which can be expanded in a power series in the supercovariant derivatives as

$$\Omega(s) = 1 + c_0(s) + c_1^\alpha(s) D_\alpha + c_2(s) D^2.$$

(3.143)

We note that higher degrees of the spinor derivatives can be reduced to the structures which are already present in Eq. (3.143) by using the rules $D_\alpha D_\beta = i\partial_{\alpha\beta} - C_{\alpha\beta} D^2$, $(D^2)^2 = \Box$ and $D_\alpha D^2 = -i\partial_{\alpha\beta} D^\beta$. The coefficient functions c_0, c_1, c_2 depend analytically on the proper time s, the superfield Ψ and its supercovariant derivatives, and the space-time derivatives $\partial_{\alpha\beta}$, which act on the delta function appearing in Eq. (3.141).

The operator $\Omega(s)$ satisfies the differential equation

$$\frac{1}{i} \frac{d\Omega}{ds} = \Omega(D^2 + \Psi).$$

(3.144)

Substituting here the explicit form for $\Omega(s)$ into Eq. (3.142), we obtain a coupled set of differential equations for the coefficient functions $c_0(s), c_1(s), c_2(s)$,

$$\frac{1}{i} \frac{dc_0}{ds} = c_0\Psi + c_2(\Box + D^2\Psi) + c_1^\alpha(D_\alpha\Psi) + \Psi,$$

$$\frac{1}{i} \frac{dc_1^\alpha}{ds} = -ic_{1\gamma}\partial^{\gamma\alpha} + c_1^\alpha\Psi + c_2 D^\alpha\Psi,$$

$$\frac{1}{i} \frac{dc_2}{ds} = c_0 + c_2\Psi + 1.$$

(3.145)

As $\Omega(s=0) = 1$, the initial conditions are $c_0(0) = c_1^\alpha(0) = c_2(0) = 0$. Since this is a linear inhomogeneous system of differential equations with constant coefficients, its solution must have the form $c_i(s) = b_i e^{i\omega s} + d_i$, where b_i and d_i are some s-independent terms. Substituting this ansatz into the Eq. (3.145), one finds for the b_i coefficients,

$$(\omega - \Psi)b_0 = b_2(\Box + D^2\Psi) + b_1^\alpha(D_\alpha\Psi),$$

$$(\omega - \Psi)b_1^\alpha = -ib_{1\beta}\partial^{\beta\alpha} + b_2 D^\alpha\Psi,$$

$$(\omega - \Psi)b_2 = b_0,$$

(3.146)

and for the d_i coefficients,

$$d_0\Psi + d_2(\Box + D^2\Psi) + d_1^\alpha(D_\alpha\Psi) + \Psi = 0,$$
$$-id_{1\gamma}\partial^{\gamma\alpha} + d_1^\alpha\Psi + d_2 D^\alpha\Psi = 0,$$
$$d_0 + d_2\Psi + 1 = 0. \tag{3.147}$$

The system (3.146), after some simplifications, implies in the following equation:

$$b_{1\gamma}\left[\left(\omega - \Psi + \frac{1}{2}\frac{D^\beta\Psi D_\beta\Psi}{(\omega - \Psi)^2 - \Box - (D^2\Psi)}\right)C^{\alpha\gamma} - i\partial^{\alpha\gamma}\right] = 0. \tag{3.148}$$

Since $b_{1\gamma} \neq 0$ (otherwise the solution is trivial), the ω's can be found requiring the 2×2 matrix $\Delta^{\gamma\alpha}$ defined as

$$\Delta^{\gamma\alpha} = \left(\omega - \Psi + \frac{1}{2}\frac{D^\beta\Psi D_\beta\Psi}{(\omega - \Psi)^2 - \Box - (D^2\Psi)}\right)C^{\alpha\gamma} - i\partial^{\alpha\gamma} \tag{3.149}$$

to have zero determinant. The corresponding equation for ω is solvable in principle, but its evaluation is technically very difficult, and we will not obtain the solution here. The corresponding results would be extremely complicated.

Let us also briefly discuss the calculation of the one-loop effective potential via the more traditional method, that is, via summation of the supergraphs. In the simplest case of the purely scalar superfield theory, we can start with the expression (3.121) and expand it in the power series in the background field $\Psi = -V''(\Phi)$:

$$\Gamma^{(1)} = \frac{i}{2}\operatorname{Tr}\ln[D^2 + \Psi] = \frac{i}{2}\operatorname{Tr}\ln(D^2) + \frac{i}{2}\operatorname{Tr}\ln[1 + \Psi D^2\Box^{-1}] =$$
$$= \frac{i}{2}\operatorname{Tr}\ln(D^2) + \frac{i}{2}\operatorname{Tr}\sum_{n=1}^{\infty}\frac{(-1)^{n+1}}{n}[\Psi\frac{D^2}{\Box}]^n, \tag{3.150}$$

where we can construct corresponding supergraphs and calculate the sum, arriving at the same result (3.130). However, in the case of the scalar superfield coupled to the gauge one the situation is much more delicate. Let us consider for example the scalar QED whose action is

$$S_m = \int d^5z\left[\frac{1}{2}(D^\alpha\Phi + i\Phi A^\alpha)(D_\alpha\bar\Phi - iA_\alpha\bar\Phi) + \frac{1}{2g^2}W^\alpha W_\alpha\right]. \tag{3.151}$$

First of all, one can find that the triple gauge-matter vertices in the one-loop diagrams arise within the following fragments given by Fig. 3.3.

In this figure, the factor $D^\beta D^\alpha + \xi D^\alpha D^\beta$ originates from the propagator (there is no essential difference between the QED and Chern-Simons theories in this context since this factor commutes with D^2). The indices α and β are the indices of gauge fields contracted into this propagator. Let us, for the sake of the convenience, refer to a vertex as to the "proper" one if it contains three and more quantum fields since such a vertex cannot be absorbed into redefinition of the propagator in the framework of

Fig. 3.3 A fragment of a
one-loop diagram containing
a triple gauge-scalar vertex

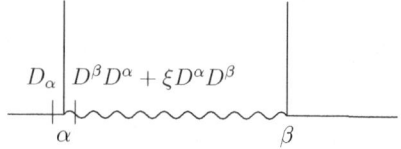

the background field approach. Otherwise, if the vertex involves only two quantum fields, it is qualified as an "improper" one. We see that if we use integration by parts to move the derivative D_α, originated from the interaction vertex, to the propagator of the gauge field proportional to $D^\beta D^\alpha + \xi D^\alpha D^\beta$, we will annihilate its gauge-independent part, and the gauge-dependent part vanishes at $\xi = 0$. Therefore we impose the gauge $\xi = 0$, an analogue of the Landau gauge [23], to simplify the calculations, i.e. to annihilate all diagrams with "improper" vertices $\Phi A^\alpha D_\alpha \bar\phi - \bar\Phi A^\alpha D_\alpha \phi$, where Φ, $\bar\Phi$ stay for external fields. As a result, all triple vertices are ruled out (the similar situation occurs in the Landau gauge also in other three-dimensional gauge theories coupled to a scalar matter), and hence all vertices in the one-loop diagrams yielding nontrivial contributions in this gauge are quartic ones, and the relevant one-loop diagrams are composed only of gauge propagators. Moreover, in the QED, the gauge propagator (3.85) involves four spinor derivatives, hence one-loop diagrams with n propagators will contain $4n$ spinor derivatives, and the expression involving $4n$ spinor derivatives turns out to be proportional to $(D^2 D^2)^n$ whose trace in the superspace is evidently equal to zero. Therefore, we have just proved that the one-loop Kählerian effective potential in the scalar super-QED, in the absence of the self-coupling of the scalar superfield, vanishes in the Landau gauge. This result has been generalized for a certain type of a higher-derivative extension of the scalar super-QED in [49].

In this section, we developed a superfield method for calculation of the effective potential in three-dimensional supersymmetric field theories. We succeeded to obtain explicit expressions for the Kählerian effective potential (which depends on the superfield Φ but not on its derivatives) up to two loops. In principle, our approach can be directly generalized for higher loops. We have also demonstrated the method for performing much more difficult calculations of non-Kählerian contributions to the effective action. The generalization for the noncommutative case is straightforward (it is clear that at the one-loop order, the results for the effective potential in commutative and noncommutative cases are the same, the difference begins with the two-loop order, see [45] for examples).

3.6 Supersymmetry in Three-Dimensional Space-Time and Noncommutativity

The concept of the space-time noncommutativity has attracted a great attention recently. One of the principal motivations for it, is the hope that introducing the noncommutative coordinates obeying the relation [50]

Fig. 3.4 The simplest tadpole (super)graph

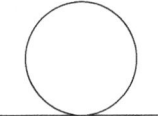

$$[x^m, x^n] = i\Theta^{mn}, \tag{3.152}$$

where Θ^{mn} is a constant matrix, can essentially improve the renormalization properties of the field theories. The most adequate formulation allowing to implement these relations within the framework of the quantum field theory [51] is based on the replacement of the usual product of the fields by their Moyal product defined as

$$\phi_1(x) * \ldots \phi_N(x) =$$

$$= \prod_{l=1}^{N} \int \frac{d^d k_i}{(2\pi)^d} (2\pi)^d e^{i(k_1 + \ldots + k_N)x} \tilde{\phi}_1(k_1) \ldots \tilde{\phi}_N(k_N) \exp(i \sum_{i<j\leq N} k_i \wedge k_j), \tag{3.153}$$

where $k_i \wedge k_j = \Theta^{mn} k_{im} k_{jn}$.

However, it turns to be that the famous UV/IR mixing effect [30] implies in the partial conversion of the ultraviolet divergences to the infrared singularities. Indeed, if one considers the d-dimensional nonsupersymmetric noncommutative ϕ^4 model, the following simplest tadpole contribution to the two-point function of the ϕ field, given by Fig. 3.4, will arise.

The contribution of this graph in d space-time dimensions is (see e.g. [52, 53])

$$S_2(p) = \frac{\lambda}{6} \int \frac{d^d k}{(2\pi)^d} \frac{2 + \cos(2k \wedge p)}{k^2 + m^2} \phi(-p)\phi(p) = -\frac{\lambda}{3} [\frac{\Gamma(1-d/2)}{(4\pi)^{d/2}(m^2)^{1-d/2}} +$$

$$+ \frac{1}{2(2\pi)^{d/2}} (m^2)^{d/2-1} \frac{K_{d/2-1}(\sqrt{m^2 \tilde{p}^2})}{(\sqrt{m^2 \tilde{p}^2})^{d/2-1}}], \tag{3.154}$$

where $\tilde{p}^m = \Theta^{mn} p_n$. We see that the first term of the expression (3.154) is similar to the common UV divergent term different from that one arising in the commutative case only by an overall numerical coefficient (this difference, from a formal viewpoint, can be explained by the fact that, when the noncommutativity is introduced, some of contractions of the fields continue to be planar ones, i.e. do not acquire the phase factor), while the second one is singular, and, since $K_n(x)|_{x\to 0} \propto \frac{1}{x^n}$, we arrive at the infrared singularity of the order $d - 2$ in the external momentum p_m, in particular, at $d = 3$ we obtain the linear singularity, and at $d = 4$—the quadratic one. A remarkable fact is that this singularity arises in the massive case as well as in the massless one where the typical momentum integral looks like

$$\int \frac{d^d k}{(2\pi)^d} \frac{\cos(2k \wedge p)}{k^2} = \frac{\Gamma(d/2 - 1)}{(4\pi)^{d/2}(\tilde{p}^2)^{d/2-1}}, \qquad (3.155)$$

which displays the same infrared singularities. Further, the infrared singularities arising due to the UV/IR mixing will be called the UV/IR infrared singularities.

The natural hope to solve this problem was related with the well-known fact that the supersymmetry improves the renormalization behaviour of the field theories eliminating some ultraviolet divergences [14]. Therefore, it is natural to expect that the situation with the infrared singularities can also be cured, at least partially. Moreover, since the commutation relation (3.152) affects only the bosonic coordinates, it is natural to expect that introducing the Moyal product into superfield theories will not affect the supersymmetry algebra, therefore the noncommutative extension of the supersymmetric field theories formulated in the superfield language will be straightforward! This idea was originally proposed in [29] for the four-dimensional Wess-Zumino model, however, it is clear that this idea can be straightforwardly applied also to the three-dimensional superspace. The first consistent three-dimensional example of the noncommutative supersymmetric field theory is the nonlinear supersymmetric sigma model studied within the component approach in [54]. It was shown in that paper that while both the $O(N)$ noncommutative nonlinear sigma model and the noncommutative $O(N)$ Gross-Neveu model within the $\frac{1}{N}$ expansion display the nonintegrable (linear) infrared singularities, the supersymmetric noncommutative nonlinear sigma model formulated within the component approach involving both these models as ingredients displays the explicit cancellation of such singularities involving only harmless logarithmic infrared divergences.

As a first example, let us study the scalar field coupled to the external gauge field. The action, for the adjoint form of the coupling, is (see e.g. [44])

$$S = \int d^5 z \left[\frac{1}{2} (D^\alpha \bar{\phi} + i[\bar{\phi}, A^\alpha]) * (D_\alpha \phi - i[A_\alpha, \phi]) + m\phi\bar{\phi} \right], \qquad (3.156)$$

which is a straightforward analogue of the theory (3.63), but with the algebraic commutators replaced by the Moyal ones. Due to the Moyal commutators, the vertices will acquire the phase factors and look like

$$V_3 = \sin(k_2 \wedge k_3) A^\alpha(k_1) (\phi(k_2) D_\alpha \bar{\phi}(k_3) - D_\alpha \phi(k_2) \bar{\phi}(k_3));$$
$$V_4 = 2\sin(k_1 \wedge k_2)\sin(k_3 \wedge k_4)\phi(k_1)A^\alpha(k_2)A_\alpha(k_3)\bar{\phi}(k_4). \qquad (3.157)$$

The corresponding supergraphs are just those ones depicted at Fig. 3.1. The only modification of our calculations with respect to those ones carried out in Sect. 3.4 will consist in arising the additional factor $4\sin^2(k \wedge p)$ in all contributions, as a result, the final expression for the two-point function of the gauge field will take the form

$$iS_3(p) = -2 \int d^2\theta \int \frac{d^3k}{(2\pi)^3} \frac{\sin^2(k \wedge p)}{(k^2 + m^2)[(k + p)^2 + m^2]} \tag{3.158}$$

$$\times (k_{\gamma\beta} + mC_{\gamma\beta}) \left[(D^2 A^\gamma(-p, \theta)) A^\beta(p, \theta) + \frac{1}{2} D^\gamma D^\alpha A_\alpha(-p, \theta) A^\beta(p, \theta) \right].$$

This expression can be shown to imply in the action similar to the Maxwell-Chern-Simons form [15], whose explicit expression is

$$iS_3(p) = \int d^2\theta f(p)(W^\alpha(-p)W_\alpha(p) + 2m W^\alpha(p)A_\alpha(p)), \tag{3.159}$$

where

$$f(p) = \int \frac{d^3k}{(2\pi)^3} \frac{\sin^2(k \wedge p)}{(k^2 + m^2)[(k + p)^2 + m^2]}.$$

This result, in the case of the noncommutative supersymmetric CP^{N-1} model [15], implies in the nontrivial effective dynamics for the originally purely external A^α field. Indeed, if we consider, instead of one scalar superfield ϕ, a set of N scalar superfields ϕ_i, after adding an appropriate gauge-fixing term, we find the following effective propagator for the A_α field:

$$\langle A_\alpha(-p, \theta_1) A_\beta(p, \theta_2) \rangle = \frac{4\pi i}{N f(p)} \tag{3.160}$$

$$\left[\frac{(D^2 - 2m) D_\beta D_\alpha}{p^2(p^2 + 4m^2)} + \xi \frac{D^2 D_\alpha D_\beta}{p^4} \right] \delta_{12}. \tag{3.161}$$

This propagator decreases as $\frac{1}{p}$, at large momenta since $f(p) \simeq \frac{\pi}{\sqrt{p^2}}$ in this limit. This implies that the CP^{N-1} theory is not finite but only renormalizable, in the lower order of $\frac{1}{N}$ expansion (and, probably, also in higher orders). The similar asymptotics of the effective propagator of the auxiliary Σ field takes place in the supersymmetric nonlinear noncommutative sigma model [42] which is renormalizable in all orders.

However, we should note that the cancellation of the linear singularities arising from these graphs (as well as from the graphs in the noncommutative supersymmetric QED, Chern-Simons and Maxwell Chern-Simons theories which will be considered below) occurs here due to the gauge symmetry rather that to the supersymmetry. Moreover, the logarithmic divergences also vanish (we note that, due to peculiarities of an odd-dimensional space, the one-loop logarithmic divergences can arise only in theories with highly unusual effective dynamics like the noncommutative sigma model [42] where one of the propagators is proportional to $\frac{1}{\sqrt{k^2}}$).

Further, the renormalizability of the three-dimensional supersymmetric noncommutative gauge theories was studied in great details [15]. Let us give a brief review on renormalizability of these theories. First, let us write down the following contributions to the two-point function of the gauge superfield. They are given by supergraphs depicted at Fig. 3.5.

These diagrams arise in the noncommutative supersymmetric QED as well as to the noncommutative supersymmetric Chern-Simons or Maxwell-Chern-Simons

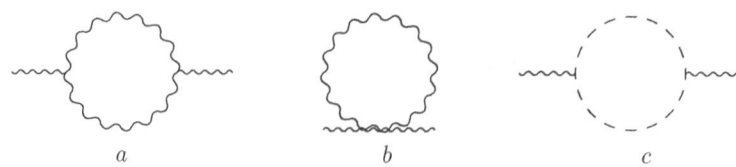

Fig. 3.5 Contributions to the two-point function of the gauge superfield from the purely gauge sector

theories [15] whose actions can be obtained from the actions (3.76) for the NC QED and (3.78) for the NC Chern-Simons theories (and their sum for the NC Maxwell-Chern-Simons theory) given in the Sect. 3.4, by replacement of the algebraic products and commutators by the Moyal ones. The same operation must be performed in the ghost action (3.79). Using the expressions for the superficial degree of divergence (3.96) and (3.97), one can show that there is no other linearly divergent graphs in the one-loop approximation as well as in higher loop orders. As we already noted above, by the symmetry reasons there is no one-loop logarithmic UV divergences, as it occurs also in usual odd-dimensional field theories.

One can find that the leading, linearly divergent parts of these contributions, for all these theories, look like

$$S_a(p) = \frac{\xi}{2} \int d^2\theta\, A^\alpha(-p) A_\alpha(p) \int \frac{d^3k}{(2\pi)^3} \frac{\sin^2(k \wedge p)}{k^2};$$

$$S_b(p) = \frac{1}{2}(1-\xi) \int d^2\theta\, A^\alpha(-p) A_\alpha(p) \int \frac{d^3k}{(2\pi)^3} \frac{\sin^2(k \wedge p)}{k^2};$$

$$S_c(p) = -\frac{1}{2} \int d^2\theta\, A^\alpha(-p) A_\alpha(p) \int \frac{d^3k}{(2\pi)^3} \frac{\sin^2(k \wedge p)}{k^2}. \qquad (3.162)$$

Sum of these contributions is equal to zero, hence the two-point function of the gauge superfield is one-loop finite and free of the UV/IR singularities. As for the possibility for logarithmic UV/IR infrared singularities, all one-loop potentially logarithmically divergent contributions are easily shown to be proportional to $\frac{\tilde{p}^m}{\sqrt{\tilde{p}^2}}$, which is actually even not a logarithmic divergence but a mild removable singularity which does not blow up. Thus, these theories are one-loop finite. At the same time, the analysis of the superficial degree of divergence shows that NC supersymmetric QED and Maxwell-Chern-Simons theories are finite in three and higher loop orders, as we noted above.

3.7   On the Noncommutativity in the Fermionic Sector of the Superspace

One of the possible extension of the noncommutativity concept is the deformation of the supersymmetry algebra carried out in the following way [55].

Let us suppose that the spinor coordinates of the superspace, instead of the Grassmann algebra, form the Clifford algebra, thus obeying the following anticommutation relations:

$$\{\theta^\alpha, \theta^\beta\} = \Sigma^{\alpha\beta}. \tag{3.163}$$

It is clear that, to maintain the key relation of the supersymmetry algebra, that is, (3.9), we should deform the supersymmetry generators, which will imply in a need to deform the spinor supercovariant derivatives via introducing second-derivative terms into them, therefore the Leibnitz rule will not be satisfied. Therefore, we must try to extend the supersymmetry (indeed, in the four-dimensional case the similar deformation of the superspace had implied in a partial breaking of the supersymmetry [31]).

In the first way, we introduce an additional set of the supersymmetry generators thus considering the extended, $\mathcal{N} = 2$ supersymmetry, with the generators are

$$Q_\alpha^i = i\partial_\alpha^i + \theta^{i\beta}\partial_{\beta\alpha}, \tag{3.164}$$

with $i = 1, 2$ is a number of the set of spinor coordinates (and, hence—of the supersymmetry generators). Then, we suppose that only in *one* of the sets of the spinor coordinates, say $i = 2$, the anticommutation relations are deformed in a manner (3.163) while for $i = 1$ they persist to be the same. Thus, only the unbroken generators Q_α^1 satisfy the usual anticommutation relation $\{Q_\alpha^1, Q_\beta^1\} = 2i\partial_{\alpha\beta}$. The supercovariant derivatives anticommuting with them are

$$D_\alpha^i = \partial_\alpha^i + i\theta^{i\beta}\partial_{\beta\alpha}, \tag{3.165}$$

with the only deformed anticommutation relation

$$\{D_\alpha^2, D_\beta^2\} = 2i\partial_{\alpha\beta} - \Sigma^{\gamma\delta}\partial_{\alpha\gamma}\partial_{\beta\delta}, \tag{3.166}$$

while all other anticommutation relations between the supercovariant derivatives persist to be the same.

The Moyal product compatible with this definition of the supersymmetry is defined as

$$\Phi_1(z) * \Phi_2(z) = \exp(-\frac{1}{2}\Sigma^{\alpha\beta}D_{\alpha 1}^2 D_{\beta 2}^2)\Phi_1(z_1)\Phi_2(z_2)|_{z_1=z_2=z}. \tag{3.167}$$

It is easy to see that in this case already the quadratic action will suffer noncommutative deformation which is a highly unusual situation (one should remind that for the usual bosonic Moyal product (3.153), the quadratic action does not suffer any deformation, while the interaction vertices are deformed).

In the second way, we propose the following generators:

$$Q_\alpha^1 = i\partial_\alpha^1 + \theta^{2\beta}\partial_{\beta\alpha},$$
$$Q_\alpha^2 = i\partial_\alpha^2 + \theta^{1\beta}\partial_{\beta\alpha}. \tag{3.168}$$

Then, we again impose the nontrivial anticommutation relation $\{\theta^{2\alpha}, \theta^{2\beta}\} = \Sigma^{\alpha\beta}$. The Moyal product is again chosen in the form (3.167). It is easy to see that in this case the quadratic action is not deformed, and the only impact of the nontrivial anticommutation relations will present in arising of additive vertices of interaction, just as in [31]. This approach has received further development in the case of $\mathcal{N} = 2$ supersymmetry where the resulting supersymmetry algebra is rather similar to the four-dimensional $\mathcal{N} = 1$ supersymmetry algebra [56]. However, studying of quantum effects within both these approaches is a completely open problem.

3.8 Conclusion

In this chapter we gave a brief introduction to the properties of three-dimensional supersymmetric field theories within the superfield approach. We described the structure of the most important three-dimensional supermultiplets, that is, scalar and spinor ones. We developed a detailed description of properties of the supercovariant derivatives and of the superfield approach for calculating quantum corrections, including the prescriptions for obtaining the effective action in one-loop and higher-loop orders. The examples we presented show that supersymmetry indeed allows to improve essentially the renormalization properties of the field theories. Really, almost all supersymmetric field theories formulated in three-dimensional space-time are one-loop finite except of the exotic theories with an effective dynamics providing a nontrivial asymptotic behavior of the effective propagators, such as, for example, the three-dimensional nonlinear sigma model [42] and CP^{N-1} model [15]. Moreover, the three-dimensional supersymmetric QED is explicitly finite in all loop orders (it was argued in [37] and explicitly proved in [41]).

This improvement of the renormalization properties plays an important role in the context of the noncommutative field theories. Indeed, it is well known that the noncommutative theories suffer from arising of nonintegrable infrared divergences due to the UV/IR mixing mechanism which converts part of ultraviolet divergences to infrared singularities whose presence in a general case can break the perturbative expansion [30]. At the same time, we have shown that the supersymmetry improves convergence of the Feynman diagrams, leaving, for many theories, only logarithmic UV divergences which in the noncommutative case are just converted to harmless logarithmic IR singularities whose presence does not break the perturbative expansion. This is the key result shown in the papers [15]. At the same time, we have observed that the formalism of the supercovariant D-algebra is applicable in the noncommutative case as well as in the commutative one, for example, the calculations of the noncommutative analogue of the supergraphs studied in Sect. 3.4, differ only by multiplying of all contributions by the common phase factor, which is equal to 1 in the case of the coupling in the fundamental representation and $\frac{1}{4}\sin^2(k \wedge p)$, where $k \wedge p = k_m \Theta^{mn} p_n$, and Θ^{mn} is a noncommutativity matrix, in the adjoint representation case (see [15] for more details).

We should note that neither the supersymmetric multiplets (scalar and spinor one) nor the actions presented in this section are unique possible ones. The important example of the action for the spinor supersymmetric multiplet is the Dirac-like action whose properties were studied in [40]. Another important example of the multiplet and the corresponding action is the three-dimensional superfield supergravity discussed in [23]. However, detailed perturbative study of this theory was not yet carried out.

Let us briefly discuss the most important actual lines of studies in the three-dimensional supersymmetric theories. Presently, the main line of interest in the three-dimensional supersymmetry is related to the extended supersymmetric Chern-Simons theories, especially, $\mathcal{N} = 6$ and $\mathcal{N} = 8$ ones, which are known to be super-conformally invariant and thus finite [4]. Another interesting line of study of these theories is based on the essentially $\mathcal{N} = 2$ supersymmetric description of such theories, whose properties are very similar to the usual $\mathcal{N} = 1$ supersymmetry in four-dimensional space-time [57]. Different aspects of the $\mathcal{N}$-extended supersymmetric Chern-Simons theories are presented in [58]. Besides of this, one more very important line of studies in these theories could consist in a detailed investigation of possible applications of three-dimensional supersymmetry within the condensed matter context where one of the first steps has been done in [24].

We close this chapter by mentioning of the noncommutative superspace where the anticommutation relations between fermionic superspace coordinates are the Clifford ones instead of the Grassmann ones. However, such a formulation seems to be realized only by paying a price of a partial supersymmetry breaking, i.e. to proceed with this formulation in the three-dimensional space one should start with the $\mathcal{N}$-extended supersymmetric theories, with further some of the supersymmetries are broken. A first attempt of developing such a formulation is presented in [55], however, this issue certainly requires more studies.

Chapter 4
Four-Dimensional Superfield Supersymmetry

Basing on our experience with studies of three-dimensional supersymmetric field theory models, now we turn to our main aim—study of a superfield description for four-dimensional supersymmetric field theories. We discuss their superfield formulations and manners to calculate quantum corrections for these theories.

4.1 General Properties of the Four-Dimensional Superspace

Now, after we have described in details the superfield approach for the supersymmetric theories formulated in the three-dimensional space-time, let us go to the usual, four-dimensional case.

The essential difference of the four-dimensional space-time from the three-dimensional one consists in the fact that the Lorentz group $SO(1, 3)$ characterizing the rotational symmetry of the space-time, possesses two mutually conjugated and linearly independent spinor representations, denoted by undotted and dotted indices respectively. Each of these representations is realized by a group of unimodular complex 2×2 matrices, that is, $SL(2, C)$, and two linear spaces where these representations are acting are the spaces of undotted and dotted spinors ψ^α, $\bar{\psi}^{\dot\alpha}$ respectively. It is easy to see that the invariant tensors of these two representations are the Levi-Civita symbols $\epsilon^{\alpha\beta}$, $\epsilon^{\dot\alpha\dot\beta}$. These tensors play the role of metrics for spinors and will be used to form the invariant scalar products of any spinors:

$$\psi^\alpha = \epsilon^{\alpha\beta}\psi_\beta; \quad \psi_\alpha = \psi^\beta \epsilon_{\beta\alpha};$$
$$\bar{\psi}_{\dot\alpha} = \epsilon_{\dot\alpha\dot\beta}\bar{\psi}^{\dot\beta}; \quad \bar{\psi}^{\dot\alpha} = \bar{\psi}_{\dot\beta}\epsilon^{\dot\beta\dot\alpha}. \tag{4.1}$$

© The Author(s), under exclusive license to Springer Nature Switzerland AG 2021
A. Petrov, *Quantum Superfield Supersymmetry*, Fundamental Theories of Physics 202,
https://doi.org/10.1007/978-3-030-68136-4_4

From a formal viewpoint, we can say that the Levi-Civita symbols "act from a definite side". These notations have been used in the book [33] and in most part of papers dealing with the four-dimensional superfield theories, and will be used henceforth; note, however, that other conventions also can be introduced, see, for example, [23]). Unlike the previous chapter, in the four-dimensional case we use the definitions of squares of spinors without $\frac{1}{2}$ factor, $\psi^2 = \psi^\alpha \psi_\alpha$, which matches the conventions used in [32, 33] and many other books and papers. It is clear that the indices $\alpha, \dot\alpha$ can take values 1,2 only. We have $\epsilon_{12} = \epsilon^{12} = \epsilon_{\dot1\dot2} = \epsilon^{\dot1\dot2} = 1$ (we note that this definition differs from that one used in the previous chapter where one had $C_{12} = -C^{12}$).

As a next step, we introduce two types of Grassmannian variables θ^α, $\bar\theta^{\dot\alpha}$ which, together with the usual bosonic coordinates x^a will play the role of coordinates on the new extended space which will be called the superspace. The spinors θ^α, $\bar\theta_{\dot\alpha}$ are mutually conjugated which is reasonable since they are transformed under mutually conjugated spinor representations of the Lorentz group. The conjugation is defined as $(\theta^\alpha)^\dagger = \bar\theta_{\dot\alpha}$, $(\theta^2)^\dagger = \bar\theta^2$. In the case of arbitrary spinors ψ^α, χ^β we can apply the following conjugation rule: $(\psi^\alpha \chi_\alpha)^\dagger = \bar\chi_{\dot\alpha} \bar\psi^{\dot\alpha}$.

Therefore we can introduce the generalized coordinates $z^A = (x^a, \theta^\alpha, \bar\theta^{\dot\alpha})$ parametrizing the superspace. In principle, it is possible to consider, instead of one set of Grassmannian coordinates $(\theta^\alpha, \bar\theta^{\dot\alpha})$, several, say $\mathcal{N}$, sets of such coordinates $(\theta^{\alpha i}, \bar\theta^i_{\dot\alpha})$, with $i = 1, \ldots \mathcal{N}$. In this case we can speak about the $\mathcal{N}$-extended supersymmetry. For $\mathcal{N} = 2$, the superfield formalism is well formulated through the methodologies of the harmonic superspace [19, 20] and of the projective one [59]. The harmonic superspace formulation is also known in $\mathcal{N} = 3$ case [60]. However, again, as in the previous chapter, we note that in most typical cases the $\mathcal{N}$-extended supersymmetric theories can be represented in terms of $\mathcal{N} = 1$ superfields, therefore we will everywhere in this chapter use the $\mathcal{N} = 1$ descriptions for all models.

The supersymmetry transformations for coordinates are

$$\delta\theta^\alpha = \epsilon^\alpha; \quad \delta\bar\theta_{\dot\alpha} = \epsilon_{\dot\alpha}; \quad \delta x^a = -\epsilon\sigma^a\bar\theta + \bar\epsilon\sigma^a\theta. \tag{4.2}$$

Here $\epsilon^\alpha, \bar\epsilon^{\dot\alpha}$ are fermionic parameters. We use the notations $\epsilon\sigma^a\bar\theta \equiv \epsilon^\alpha\sigma^a_{\alpha\dot\alpha}\bar\theta^{\dot\alpha}$ and $\bar\epsilon\bar\sigma^a\theta \equiv \bar\epsilon^{\dot\alpha}\bar\sigma^a_{\dot\alpha\alpha}\theta^\alpha$. Just as in the previous chapter, we can introduce the bispinor notation for the usual space-time derivatives: $\partial_{\alpha\dot\alpha} = \sigma^a_{\alpha\dot\alpha}\partial_a$.

The superfield is defined as a generic function of the superspace coordinates. We suggest it to have the form of a power series in spinor superspace coordinates $(\theta^\alpha, \bar\theta^{\dot\alpha})$, which allows to present it in the form of the expansion (cf. e.g. [32, 61]):

$$\mathcal{F}(x, \theta, \bar\theta) = A(x) + \theta^\alpha\psi_\alpha(x) + \bar\theta_{\dot\alpha}\zeta^{\dot\alpha}(x) + \theta^2 F(x) + \bar\theta^2 G(x) + i(\bar\theta\sigma^a\theta)A_a(x) +$$
$$+ \bar\theta^2\theta^\alpha\chi_\alpha(x) + \theta^2\bar\theta_{\dot\alpha}\xi^{\dot\alpha}(x) + \theta^2\bar\theta^2 H(x). \tag{4.3}$$

We note that this power series is finite due to anticommuting property of Grassmann spinor coordinates $\theta, \bar\theta$ which immediately annihilates third and higher degrees in $\theta, \bar\theta$. Further we will see that there are some restrictions on structure of superfields caused by the form of a representation of the supersymmetry algebra. Here $f(x), \psi_\alpha(x), \ldots$

are bosonic and fermionic fields forming a component content of the superfield $\mathcal{F}$ given by the expression above. We note that, just as in the three-dimensional case, the numbers of bosonic and fermionic degrees of freedom in any supersymmetric theory are equal. If a theory describing dynamics of these fields is supersymmetric, its action should be invariant under supersymmetry transformations which are defined as symmetry transformations with fermionic parameters.

Example. The Wess-Zumino model [32], whose action, in the simplest case, that is, the free massless theory, looks like

$$S = \int d^4x (\bar{\phi}\Box\phi - \frac{i}{2}\bar{\psi}^{\dot{\alpha}}\partial_{\dot{\alpha}\alpha}\psi^\alpha + \bar{F}F), \tag{4.4}$$

is invariant under the following transformations

$$
\begin{aligned}
\delta\phi(x) &= \epsilon^\alpha\psi_\alpha(x); \\
\delta\psi_\alpha(x) &= \epsilon_\alpha\bar{F}(x) - \bar{\epsilon}^{\dot{\alpha}}i\partial_{\alpha\dot{\alpha}}\phi(x); \\
\delta F(x) &= \bar{\epsilon}_{\dot{\alpha}}i\partial^{\alpha\dot{\alpha}}\psi_\alpha,
\end{aligned} \tag{4.5}
$$

with the analogous transformation for the conjugated fields $\bar{\phi}(x)$, $\bar{\psi}_{\dot{\alpha}}(x)$, $\bar{F}(x)$. Since ϵ^α, $\bar{\epsilon}^{\dot{\alpha}}$ are the (global) fermionic parameters, these transformations are the supersymmetry ones. We see that variations of bosonic fields under these transformations are proportional to fermionic fields and vice versa. The Wess-Zumino model will be discussed in details in the next section.

Now, let us require that the variation of an arbitrary superfield $\mathcal{F}(x, \theta, \bar{\theta})$ under the supersymmetry transformations has the form similar to usual translations, just as in the three-dimensional case, i.e.

$$\delta\mathcal{F}(x, \theta, \bar{\theta}) = (\epsilon^\alpha Q_\alpha + \bar{\epsilon}_{\dot{\alpha}}\bar{Q}^{\dot{\alpha}})\mathcal{F}(x, \theta, \bar{\theta}). \tag{4.6}$$

Here we suppose that Q_α, $\bar{Q}_{\dot{\alpha}}$ are generators of supersymmetry obeying (anti) commutation relations

$$\{Q_\alpha, \bar{Q}_{\dot{\alpha}}\} = 2i\sigma^m_{\alpha\dot{\alpha}}\partial_m; \ \{Q_\alpha, Q_\beta\} = \{\bar{Q}_{\dot{\alpha}}, \bar{Q}_{\dot{\beta}}\} = 0; \ [Q_\alpha, \partial_m] = 0. \tag{4.7}$$

The ϵ^α, $\bar{\epsilon}^{\dot{\alpha}}$ are the infinitesimal global parameters of the supersymmetry transformation. It is natural to treat the variation (4.6) as a translation in the superspace characterized by coordinates $(x^a, \theta^\alpha, \bar{\theta}_{\dot{\alpha}})$ (to be more precise, this is a translation in the fermionic sector of the superspace). This form of anticommutators is the simplest one allowing for nontrivial unification of the supersymmetry algebra with the Poincaré algebra.

Translations in the superspace are presented by standard Poincaré translations and transformations (4.6) which can be called supertranslations. It is easy to see that (4.6) is a manifestly Lorentz covariant transformation. As we already have noted, the simplest, $\mathcal{N} = 1$ superspace is parametrized by 4 bosonic coordinates x^a and

4 fermionic ones θ^α, $\bar{\theta}^{\dot\alpha}$ so it is 8-dimensional and will be denoted henceforth as $R^{4|4}$. We will refer to it as to the four-dimensional superspace. It is natural to assume that the superfields of the forms like (4.3), in the new supersymmetric field theory, will be dynamical variables playing the role of fields in the superspace. The task of this chapter consists in development of the quantum theory for superfields in the four-dimensional superspace basing on principles of the standard methodology of quantum field theory.

We introduce derivatives on the superspace in a manner similar to the three-dimensional case, that is, we use the same definition of the (left) Grassmannian derivative with respect to θ^α (3.1), and the derivative with respect to $\bar{\theta}_{\dot\alpha}$ is defined analogously.

Then, to introduce the integral we start with the usual Grassmannian definition $\int d\theta\theta = 1$, which we generalize to

$$\int d\theta_\alpha \theta^\beta = \delta_\alpha^\beta.$$

It is a convention. Just as in three-dimensional theories, the θ and $d\theta$ have different (and opposite) mass dimensions—again, the dimension of θ is equal to $-\frac{1}{2}$, and of $d\theta$—to $\frac{1}{2}$, and variation $\delta\theta$ never should be mixed with differential $d\theta$ since they have different dimensions. Then, an integral from a constant is zero,

$$\int d\theta \cdot 1 = 0,$$

this identity is caused by suggestion of the translation invariance due to which the relation $\int d\theta(\theta + \lambda) = \int d\theta\theta$ for constant λ must be satisfied, hence $\lambda \int d\theta = 0$. We introduce the following scalar measures for Grassmann integration:

$$d^2\theta = -\frac{1}{4}d\theta^\alpha d\theta_\alpha, \ d^2\bar{\theta} = -\frac{1}{4}d\bar{\theta}_{\dot\alpha}d\bar{\theta}^{\dot\alpha}, \ d^4\theta = d^2\theta d^2\bar{\theta}. \tag{4.8}$$

These measures satisfy the relations

$$\int d^2\theta\theta^2 = \int d^2\bar{\theta}\bar{\theta}^2 = \int d^4\theta\theta^2\bar{\theta}^2 = 1. \tag{4.9}$$

Since $\frac{\partial\theta^\alpha}{\partial\theta^\beta} \equiv \partial_\alpha\theta^\beta = \delta_\beta^\alpha$ as well as $\int d\theta_\alpha\theta^\beta = \delta_\alpha^\beta$ (similarly, $\partial^{\dot\alpha}\bar{\theta}_{\dot\beta} = \int d\bar{\theta}^{\dot\alpha}\bar{\theta}_{\dot\beta} = \delta_{\dot\beta}^{\dot\alpha}$) we conclude that integration and differentiation in Grassmannian space described by $(\theta^\alpha, \bar{\theta}_{\dot\alpha})$ are equivalent, similarly to the three-dimensional case. In particular, we see that

$$\int d^4\theta F(x, \theta, \bar{\theta}) = \frac{1}{16}\frac{\partial^2}{\partial\theta^2}\frac{\partial^2}{\partial\bar{\theta}^2}F(x, \theta, \bar{\theta}) = \frac{1}{16}F(x, \theta, \bar{\theta})|_{\theta^2\bar{\theta}^2};$$

$$\int d^2\theta G(x, \theta) = -\frac{1}{4}\frac{\partial^2}{\partial\theta^2}G(x, \theta) = -\frac{1}{4}G(x, \theta)|_{\theta^2}. \tag{4.10}$$

Here $|_{\theta^2}$, $|_{\theta^2\bar{\theta}^2}$ denotes the corresponding component of the superfield. Of course, differentiations with respect to Grassmannian coordinates anticommute.

The supersymmetry generators are required to satisfy the anticommutation relations (4.7) and possess several realizations in terms of $\frac{\partial}{\partial x^m}$ and $\frac{\partial}{\partial\theta_\alpha}$, $\frac{\partial}{\partial\bar{\theta}_{\dot{\alpha}}}$, e.g.

$$\bar{Q}_{\dot{\alpha}} = i(\frac{\partial}{\partial\bar{\theta}^{\dot{\alpha}}} - i\theta^\alpha(\sigma^m)_{\alpha\dot{\alpha}}\partial_m), \quad Q_\alpha = i(\frac{\partial}{\partial\theta^\alpha} + i\bar{\theta}^{\dot{\beta}}(\bar{\sigma}^m)_{\dot{\beta}\alpha}\partial_m). \tag{4.11}$$

This realization, or, as is the same, this representation of the supersymmetry algebra is not unique, other forms of these generators can be introduced as well. However, we emphasize again that all its possible representations must satisfy the relations (4.7).

The spinor supercovariant derivatives D_A also must be constructed from $\frac{\partial}{\partial x^m}$ and $\frac{\partial}{\partial\theta_\alpha}$, $\frac{\partial}{\partial\bar{\theta}_{\dot{\alpha}}}$. They should anticommute with generators Q_α, $\bar{Q}_{\dot{\alpha}}$ which provides that $D_A\Phi$ is transformed covariantly, i.e. according to (4.6), for any superfield Φ we can write

$$D_A\delta\Phi = \delta(D_A\Phi) = (\epsilon Q + \bar{\epsilon}\bar{Q})D_A\Phi,$$

which implies that $\{D_\alpha, Q_\beta\} = \{\bar{D}_{\dot{\alpha}}, Q_\beta\} = \{D_\alpha, \bar{Q}_{\dot{\beta}}\} = \{\bar{D}_{\dot{\alpha}}, \bar{Q}_{\dot{\beta}}\} = 0$. For example, if generators of the supersymmetry are chosen in the form (4.11), the corresponding supercovariant derivatives are realized as

$$\bar{D}_{\dot{\alpha}} = -i\bar{Q}_{\dot{\alpha}} + 2i\theta^\alpha\partial_{\alpha\dot{\alpha}} = \frac{\partial}{\partial\bar{\theta}^{\dot{\alpha}}} + i\theta^\alpha(\sigma^m)_{\alpha\dot{\alpha}}\partial_m,$$

$$D_\alpha = -iQ_\alpha - 2i\bar{\theta}^{\dot{\alpha}}\partial_{\alpha\dot{\alpha}} = \frac{\partial}{\partial\theta^\alpha} - i\bar{\theta}^{\dot{\beta}}(\bar{\sigma}^m)_{\dot{\beta}\alpha}\partial_m. \tag{4.12}$$

The spinor supercovariant derivatives satisfy the following anticommutation relations

$$\{D_\alpha, \bar{D}_{\dot{\alpha}}\} = 2i\partial_{\alpha\dot{\alpha}}; \quad \{D_\alpha, D_\beta\} = \{\bar{D}_{\dot{\alpha}}, \bar{D}_{\dot{\beta}}\} = 0. \tag{4.13}$$

So we defined the operations of integration and differentiation in the superspace. Further we will use the integral measure for the complete superspace $d^8z = d^4xd^2\theta d^2\bar{\theta}$, the integral measure for the chiral subspace $d^6z = d^4xd^2\theta$ and the integral measure for the antichiral one $d^6\bar{z} = d^4xd^2\bar{\theta}$. We also can use the identities $D^2\theta^2 = \bar{D}^2\bar{\theta}^2 = -4$.

Now, let us introduce the Grassmannian delta function. We suggest that it must satisfy the condition analogous to that one for a standard delta function

$$\int d^4\theta'\delta^4(\theta - \theta')f(\theta') = f(\theta). \tag{4.14}$$

This identity can be satisfied if we choose

$$\delta^4(\theta - \theta') = (\theta - \theta')^2(\bar{\theta} - \bar{\theta}')^2. \tag{4.15}$$

It is easy to see that this delta function fulfils the condition

$$\int d^4\theta \, \delta^4(\theta - \theta') = 1. \tag{4.16}$$

From now, we will use the convenient notation $\delta_{12} \equiv \delta^4(\theta_1 - \theta_2)$.

For the Grassmannian delta functions, there is a fundamental identity

$$\delta_{12} D_1^2 \bar{D}_2^2 \delta_{12} = 16\delta_{12}. \tag{4.17}$$

Further, it will be used within perturbative calculations, to shrink a loop into a point in the Grassmannian space. To prove (4.17), one can use the expansion of supercovariant derivatives (4.12) and note that due to the evident property $\delta_{12}\frac{\partial}{\partial\theta^\alpha}\delta_{12} = \frac{\partial}{\partial\theta^\alpha}\delta_{12}|_{\theta_1=\theta_2} = 2(\theta_{1\alpha} - \theta_{2\alpha})(\bar{\theta}_1 - \bar{\theta}_2)^2|_{\theta_1=\theta_2} = 0$, and some other similar relations, only terms of the form $\delta_{12}(\frac{\partial}{\partial\theta})^2(\frac{\partial}{\partial\theta})^2\delta_{12} = 16\delta_{12}$ survive. It is easy to see that $\delta_{12}\delta_{12} = \delta_{12}D^\alpha\delta_{12} = \delta_{12}D^2\delta_{12} = \delta_{12}\bar{D}_{\dot\alpha}\delta_{12} = 0$.

A supermatrix is defined as a matrix $M = M_Q^P$ of the form

$$M = \begin{pmatrix} A & B \\ C & D \end{pmatrix}. \tag{4.18}$$

determining a quadratic form $z_P M_Q^P z'^Q$ with z, z' are coordinates on the superspace. Here A, B, C, D are even-even, even-odd, odd-even and odd-odd blocks respectively. Superdeterminant of this matrix is introduced as

$$\mathrm{sdet}\,M = \int d^8 z_1 d^8 z_2 \exp(-z_1 M z_2). \tag{4.19}$$

It is equal to

$$\mathrm{sdet}\,M = \det A \det^{-1}(D - CA^{-1}B). \tag{4.20}$$

And a supertrace is equal to $\mathrm{Str}\,M = \sum_A (-1)^{\epsilon_A} M_A^A = \mathrm{tr}\,A - \mathrm{tr}\,D$. As usual, $\mathrm{sdet}\,M = \exp(\mathrm{Str}\,\log M)$.

We can introduce change of variables in superspace. So, if the coordinates are changed as

$$x'^a = x'^a(x, \theta, \bar{\theta}); \ \theta'^\alpha = \theta'^\alpha(x, \theta, \bar{\theta}), \ \bar{\theta}'^{\dot\alpha} = \bar{\theta}'^{\dot\alpha}(x, \theta, \bar{\theta}), \tag{4.21}$$

the measure of integral over the superspace is transformed as

$$d^4x'd^4\theta' = d^4x d^4\theta \; \mathrm{sdet}(\frac{\partial z'}{\partial z}), \qquad (4.22)$$

where supermatrix $(\frac{\partial z'}{\partial z})$ is

$$\frac{\partial z'}{\partial z} = \begin{pmatrix} \frac{\partial x'}{\partial x} & \frac{\partial x'}{\partial \theta} & \frac{\partial x'}{\partial \bar\theta} \\ \frac{\partial \theta'}{\partial x} & \frac{\partial \theta'}{\partial \theta} & \frac{\partial \theta'}{\partial \bar\theta} \\ \frac{\partial \bar\theta'}{\partial x} & \frac{\partial \bar\theta'}{\partial \theta} & \frac{\partial \bar\theta'}{\partial \bar\theta} \end{pmatrix}. \qquad (4.23)$$

Now, let us define two most used superfields in $\mathcal{N} = 1$ superspace. The first of them is the real scalar superfield whose form is given by (4.3), with the additional condition $V^\dagger = V$ (cf. [32, 43]), i.e.

$$V(x,\theta,\bar\theta) = C(x) + \theta^\alpha \chi_\alpha(x) + \bar\theta_{\dot\alpha}\bar\chi^{\dot\alpha}(x) - \theta^2 M(x) - \bar\theta^2 \bar M(x) + i(\bar\theta\sigma^a\theta)A_a(x) +$$
$$+ \, \bar\theta^2\theta^\alpha \lambda_\alpha(x) + \theta^2\bar\theta_{\dot\alpha}\bar\lambda^{\dot\alpha}(x) + \theta^2\bar\theta^2 \mathcal{D}(x). \qquad (4.24)$$

The second one is the chiral superfield. It is defined in the following way: the superfield $\Phi(z)$ is called chiral if and only if it satisfies the condition $\bar D_{\dot\alpha}\Phi = 0$. The choice of supercovariant derivatives in the form $\bar D_{\dot\alpha} = \frac{\partial}{\partial\bar\theta^{\dot\alpha}}$, $D_\alpha = \frac{\partial}{\partial\theta^\alpha} - 2i\bar\theta^{\dot\beta}(\bar\sigma^m)_{\dot\beta\alpha}\partial_m$, consistent with the anticommutation relations (4.13), allows one to reduce this condition to $\frac{\partial}{\partial\bar\theta_{\dot\alpha}}\Phi = 0$, i.e. the chiral superfield Φ turns out to be $\bar\theta$-independent which allows to represent it in the following simplest form (see also [32, 43]):

$$\Phi(x,\theta) = \phi(x) + \theta^\alpha \psi_\alpha(x) - \theta^2 F(x). \qquad (4.25)$$

At the same time, one should introduce the antichiral superfield $\bar\Phi$ which is defined to satisfy the "conjugated" condition $D_\alpha\bar\Phi = 0$. But, since the spinor supercovariant derivative D_α in the form (4.12) is *not* a straightforward analogue of the derivative $\bar D_{\dot\alpha}$, one *cannot* use the analogue of the expansion (4.25), obtained by simple replacement of fields and spinor coordinates by conjugated ones, for the antichiral field. Its form turns out to be more complicated, and besides of the straightforward analogues of those ones given in Eq. (4.25), it involves additional terms:

$$\bar\Phi(x,\theta,\bar\theta) = \bar\phi(x) + \bar\theta_{\dot\alpha}\bar\psi^{\dot\alpha}(x) - \bar\theta^2\bar F(x) + \dots, \qquad (4.26)$$

where dots are for the θ-dependent terms. At the same time, one can use the definition in terms of projections for the components of Φ:

$$\phi(x) = \Phi(z)|;$$
$$\psi_\alpha(x) = D_\alpha\Phi(z)|;$$
$$F(x) = \frac{1}{4}D^2\Phi(z))|, \qquad (4.27)$$

and the definitions for the components of $\bar{\Phi}$ can be obtained via straightforward conjugation of these definitions. Here, similarly to the previous chapter, we use the notation $\Phi(z)| \equiv \Phi(z)|_{\theta=\bar{\theta}=0}$, with the condition $\theta = \bar{\theta} = 0$ is imposed after the differentiation.

One can also define the components of the real scalar superfield in terms of projections:

$$
\begin{aligned}
C(x) &= V(z)|, \\
\chi_\alpha(x) &= D_\alpha V(z)|; \quad \bar{\chi}^{\dot{\alpha}}(x) = \bar{D}^{\dot{\alpha}} V(z)|; \\
M(x) &= \frac{1}{4} D^2 V(z))|; \quad \bar{M}(x) = \frac{1}{4} \bar{D}^2 V(z)|; \\
A_{\alpha\dot{\alpha}} &= \frac{i}{2}[D_\alpha, \bar{D}_{\dot{\alpha}}] V(z)|; \\
\lambda_\alpha(x) &= -\frac{1}{4} \bar{D}^2 D_\alpha V(z)|; \quad \bar{\lambda}^{\dot{\alpha}}(x) = -\frac{1}{4} D^2 \bar{D}^{\dot{\alpha}} V(z)|; \\
\mathcal{D}(x) &= \frac{1}{16} D^\alpha \bar{D}^2 D_\alpha V(z)|.
\end{aligned}
\tag{4.28}
$$

To develop the functional integral approach, we also must introduce the variational derivative. In an usual field theory it is defined as

$$
\frac{\delta}{\delta A(x)} \int d^4 y f(y) A(y) = f(x),
\tag{4.29}
$$

if $f(x)$ and $A(x)$ are functionally independent. Just an analogous definition can be introduced for a general (non-chiral) superfield:

$$
\frac{\delta}{\delta V(z)} \int d^8 z' f(z') V(z') = f(z).
\tag{4.30}
$$

Now, let us introduce the variational derivative with respect to a chiral superfield. As the chiral superfield depends effectively only on one set of the Grassmannian coordinates, that is, θ^α, cf. (4.25), the integral from a chiral function is non-trivial only if it is taken over the chiral subspace, i.e. over $d^6 z = d^4 x d^2\theta$. Hence we must define a variational derivative with respect to a chiral superfield Φ as

$$
\frac{\delta}{\delta \Phi(z)} \int d^6 z' F(z') \Phi(z') = F(z).
\tag{4.31}
$$

And, it follows straightforwardly from the equivalence between differentiation and integration with respect to $\theta_\alpha, \bar{\theta}_{\dot{\alpha}}$, that $\int d^8 z = \int d^6 z(-\frac{1}{4}\bar{D}^2) = \int d^6\bar{z}(-\frac{1}{4}D^2)$, thus, the variational derivative from an integral over the whole superspace with respect to a chiral superfield can be introduced as

$$\frac{\delta}{\delta\Phi(z)}\int d^8z' G(z')\Phi(z') = \frac{\delta}{\delta\Phi(z)}\int d^6z'(-\frac{1}{4}\bar{D}^2)G(z')\Phi(z') = -\frac{1}{4}\bar{D}^2 G(z).$$

(4.32)

Therefore we have $\frac{\delta\Phi(z)}{\delta\Phi(z')} = \delta_+(z-z')$ where $\delta_+(z-z') = -\frac{1}{4}\bar{D}^2\delta^8(z-z')$ is the chiral delta function. It allows us to obtain the useful relation

$$\frac{\delta^2}{\delta\Phi(z_1)\delta\bar{\Phi}(z_2)}\int d^8z\,\Phi\bar{\Phi} = \frac{1}{16}\bar{D}_1^2 D_2^2\delta^8(z_1-z_2) =$$
$$= (-\frac{1}{4})D^2\delta_+(z_1-z_2) = (-\frac{1}{4})\bar{D}^2\delta_-(z_1-z_2).$$

(4.33)

Here $\delta_-(z_1-z_2) = -\frac{1}{4}D^2\delta^8(z_1-z_2)$ is the antichiral delta function. One should also note the important identity $D_1^2\delta^8(z_1-z_2) = D_2^2\delta^8(z_1-z_2)$.

If we consider some differential operator Δ acting on superfields, we can introduce its functional supertrace and superdeterminant:

$$\mathrm{Str}\Delta = \int d^8z_1 d^8z_2 \delta^8(z_1-z_2)\Delta\delta^8(z_1-z_2).$$

(4.34)

If we introduce a kernel of the Δ which has the form $\Delta(z_1, z_2)$, we can write

$$\mathrm{Str}\Delta = \int d^8z\Delta(z, z).$$

(4.35)

The superdeterminant is defined as

$$\mathrm{sdet}\Delta = \exp\mathrm{Str}(\log\Delta).$$

(4.36)

Further we will be generally interested in theories describing dynamics of chiral and real scalar superfields. We note that the irreducible representation of the supersymmetry algebra is realized namely on these superfields [32]. Among interesting examples of such theories, there are Wess-Zumino model, general chiral superfield theory [62], SYM theory and four-dimensional dilaton supergravity [63]. All of them are constructed on the base of chiral (and antichiral) and real scalar superfields. In this chapter we consider applications of the superfield approach to these models.

4.2 Field Theory Models in the Four-Dimensional Superspace

In this section we will present some typical field theory models in the four-dimensional superspace. The key ingredients of these models are exactly the chiral and real scalar superfields introduced above. Their component contents are given by Eqs. (4.25), (4.24) respectively.

4.2.1 Chiral Superfield Models

The simplest superfield model is the Wess-Zumino model [8, 11, 32] describing the dynamics of the chiral superfield Φ. Its explicit form in the components, in the particular (free massless) case is given by the expression (4.4). Now, let us write down its superfield action. It has the form

$$S = \int d^8 z \, \Phi \bar{\Phi} + [\int d^6 z (\frac{m}{2} \Phi^2 + \frac{\lambda}{3!} \Phi^3) + h.c.] \tag{4.37}$$

First, let us briefly discuss the question of dimensions of the superfields. It is well known that the mass dimension of the coordinate x^a is -1, and hence of the spatial derivative ∂_a is 1. We have already argued in the previous section that the dimensions of the derivatives ∂_α, D_α and of the integral measure $d\theta^\alpha$ are equal to $1/2$, and of the θ^α itself—to $-1/2$ (the dimensions of the conjugated coordinates $\bar{\theta}_{\dot{\alpha}}$ and the corresponding derivatives $\partial^{\dot{\alpha}}$, $\bar{D}^{\dot{\alpha}}$ are respectively the same). All this allows us to conclude that the dimension of the superfield Φ is equal to 1, hence the coupling constant λ is dimensionless. Applying the usual argumentation of quantum field theory, we can conclude that the Wess-Zumino model is renormalizable.

Second, let us obtain the component structure of the whole Wess-Zumino action. As we have already noted, the component structure of the first term of (4.37), that is, $\int d^8 z \, \Phi \bar{\Phi}$, is given by (4.4). The component structure of the term corresponding to the integral over chiral subspace can be easily obtained:

$$\int d^6 z (\frac{m}{2} \Phi^2 + \frac{\lambda}{3!} \Phi^3) = (-\frac{1}{4}) \int d^4 x (\frac{m}{2} D^2 \Phi^2 + \frac{\lambda}{3!} D^2 \Phi^3)|, \tag{4.38}$$

then we employ the expressions (4.27) and find

$$\int d^6 z (\frac{m}{2} \Phi^2 + \frac{\lambda}{3!} \Phi^3) = \int d^4 x [-\frac{1}{4}(m + \lambda \phi) \psi^\alpha \psi_\alpha - F(m\phi + \frac{\lambda}{2} \phi^2)]. \tag{4.39}$$

This allows us to write down the complete Wess-Zumino action in components:

$$S = \int d^4 x \Big[\bar{\phi} \Box \phi - \frac{i}{2} \bar{\psi}^{\dot{\alpha}} \partial_{\dot{\alpha}\alpha} \psi^\alpha + \bar{F} F -$$
$$- (\frac{1}{4}(m + \lambda \phi) \psi^\alpha \psi_\alpha + F(m\phi + \frac{\lambda}{2} \phi^2) + h.c.) \Big]. \tag{4.40}$$

One can see that the field F has no dynamics even after adding the interaction term, therefore it is called the auxiliary field, the similar situation occurs in the three-dimensional scalar superfield theory, see Sect. 3.2. Eliminating the field F with use of its equation of motion

$$F - (m\bar{\phi} + \frac{\lambda}{2}\bar{\phi}^2) = 0 \tag{4.41}$$

implies in the theory with the ϕ^4 interaction, by this reason the Wess-Zumino model is treated as a supersymmetric extension of the ϕ^4 theory.

It is instructive to calculate the numbers of bosonic and fermionic degrees of freedom in the Wess-Zumino model. It follows from (4.40) that in this model there are four bosonic degrees of freedom (which correspond to two complex scalar fields ϕ and F) and four fermionic ones (which corresponds to two components of the complex spinor ψ_α). This confirms again that in any supersymmetric theory, numbers of bosonic and fermionic degrees of freedom are equal. Also, we note that masses of all component fields of the supermultiplet are equal. As we argued in the previous chapter, the same situation takes place in three-dimensional supersymmetric theories.

We note that the Wess-Zumino model is not an unique theory describing the quantum dynamics of the chiral superfield. Other important examples are the higher-derivative chiral superfield theories (in particular, dilaton supergravity) and general chiral superfield theories. The examples of actions of these theories in superfield form are respectively

$$S = \int d^8z\, \bar{\Phi}(\Box - M^2)\Phi + (\int d^6z\, W(\Phi) + h.c.) \tag{4.42}$$

and

$$S = \int d^8z\, K(\Phi, \bar{\Phi}) + (\int d^6z\, W(\Phi) + h.c.). \tag{4.43}$$

Here $W(\Phi)$ is a holomorphic function of the chiral superfield (or of the set of chiral superfields) but not on its derivatives, and $K(\Phi, \bar{\Phi})$ is a real function of chiral and antichiral superfields. The component forms of these theories can be obtained in the same way as above, for example, the component expression for the higher-derivative chiral superfield action [1, 63] looks like

$$S = \int d^4x\Big[\bar{\phi}\Box(\Box - M^2)\phi - \frac{i}{2}\bar{\psi}^{\dot{\alpha}}\partial_{\dot{\alpha}\alpha}(\Box - M^2)\psi^\alpha + \bar{F}(\Box - M^2)F -$$
$$- \frac{1}{4}(4W_\phi F + W_{\phi\phi}\psi^\alpha\psi_\alpha + h.c.)\Big]. \tag{4.44}$$

and for the general chiral superfield model [62]

$$S = \int d^4x\Big[- K_{\phi\bar{\phi}}(\partial^a\phi\partial_a\bar{\phi} - \bar{F}F - \frac{i}{2}\bar{\psi}^{\dot{\alpha}}\partial_{\dot{\alpha}\alpha}\psi^\alpha) -$$
$$- \frac{1}{4}(K_{\phi\phi\bar{\phi}}(F\psi^\alpha\psi_\alpha + i\partial_a\phi\bar{\psi}^{\dot{\alpha}}\partial_{\dot{\alpha}\alpha}\psi^\alpha) + h.c.) + \frac{1}{16}K_{\phi\phi\bar{\phi}\bar{\phi}}\psi^\alpha\psi_\alpha\bar{\psi}_{\dot{\alpha}}\bar{\psi}^{\dot{\alpha}} -$$
$$- \frac{1}{4}(4W_\phi F + W_{\phi\phi}\psi^\alpha\psi_\alpha + h.c.)\Big]. \tag{4.45}$$

Here $K_\phi = \frac{\partial K(\Phi,\bar{\Phi})}{\partial \Phi}|_{\Phi=\phi,\bar{\Phi}=\bar{\phi}}$ etc., the similar definitions are applied to derivatives of W, i.e. all these functions depend on the scalar fields ϕ, $\bar{\phi}$ only. We note that in the higher-derivative theory (4.44) the auxiliary field F acquires a nontrivial dynamics. The consequences of this fact, especially within the context of the spontaneous supersymmetry breaking, are discussed in [64].

However, the theory (4.44), due to the presence of higher derivatives, involves ghost states (a possible way for their eliminating is discussed in [64]), and the theory (4.45) is non-renormalizable for general forms of K and W. We will study perturbative aspects of these theories within the superfield approach further.

4.2.2   Abelian Gauge Superfield Model

Now, let us go to the real scalar superfield case. The key feature of this superfield consists in the fact that it allows for introducing the gauge symmetry on the superspace.

Indeed, let us consider the action of the real scalar superfield V (see e.g. [32]):

$$S = \frac{1}{2} \int d^8 z \, V \left(\frac{D^\alpha \bar{D}^2 D_\alpha}{8} \right) V. \tag{4.46}$$

This action is evidently gauge invariant with respect to the transformations:

$$V \to V + i(\Lambda - \bar{\Lambda}), \tag{4.47}$$

where Λ is a chiral superfield parameter, and $\bar{\Lambda}$ is an antichiral one (to show the invariance, we use the fact that $\bar{D}^2 D_\alpha \Lambda = 0$ for a chiral Λ). Further, to develop a consistent quantum description, we must fix the gauge.

This action is constructed with use of the operator $-\frac{D^\alpha \bar{D}^2 D_\alpha}{8} \equiv \Pi_{1/2}\Box$, where $\Pi_{1/2} = -\frac{D^\alpha \bar{D}^2 D_\alpha}{8\Box}$ is a projecting operator possessing the property $\Pi_{1/2}^n = \Pi_{1/2}$ for any integer $n \geq 1$. One can see that there are two projecting operators more, $\Pi_{0+} = \frac{\bar{D}^2 D^2}{16\Box}$ and $\Pi_{0-} = \frac{D^2 \bar{D}^2}{16\Box}$ satisfying the similar properties, i.e. $\Pi_{0\pm}^n = \Pi_{0\pm}$. One can say that Π_{0+} is a projector on the chiral space, Π_{0-} is a projector on the antichiral space since acting of Π_{0+} on any superfield produces the chiral superfield, and of Π_{0-}—the antichiral superfield. The $\Pi_{1/2}$ projects on the so-called linear space since, for any superfield Σ, the new superfield $\Pi_{1/2}\Sigma \equiv \tilde{\Sigma}$ possesses the properties $D^2\tilde{\Sigma} = \bar{D}^2\tilde{\Sigma} = 0$, and such a superfield is called the linear superfield. However, it is used very rarely, some results for it can be found in [23], and its interesting application for formulating the Goldstino model is presented in [65].

It is straightforward to show that the projecting operators $\Pi_{1/2}$, Π_{0+} and Π_{0-} satisfy the properties

$$\Pi_{1/2} + \Pi_{0+} + \Pi_{0-} = 1;$$
$$\Pi_{0+}\Pi_{0-} = \Pi_{0\pm}\Pi_{1/2} = \Pi_{0-}\Pi_{0+} = \Pi_{1/2}\Pi_{0\pm} = 0, \tag{4.48}$$

i.e. they form the complete and orthogonal set of the projecting operators. It implies that any superfield can be represented as a linear combination of chiral, antichiral and linear ones.

Now let us obtain the component structure of the action (4.46). In principle, it can be done via a straightforward use of the expression (4.46) and the projections (4.28). However, such a way is very cumbersome. Therefore we use another manner: since the chiral and complete measures are related by the rule $\int d^8 z = \int d^6 z (-\frac{\bar{D}^2}{4})$, we can rewrite the action (4.46) in an equivalent form (cf. [66]):

$$S = \frac{1}{64} \int d^6 z \, W^\alpha W_\alpha, \tag{4.49}$$

where

$$W_\alpha = -\bar{D}^2 D_\alpha V \tag{4.50}$$

is an Abelian superfield strength. First, it is clear that W_α is chiral. Second, one can obtain its component expansion (in the chiral representation):

$$W_\alpha = 4[\lambda_\alpha + \theta^\beta f_{\beta\alpha} + \theta_\alpha \mathcal{D} - \frac{i}{2}\theta^2 \partial_{\alpha\dot\beta}\bar\lambda^{\dot\beta}]. \tag{4.51}$$

or, in the form of projections,

$$W_\alpha| = 4\lambda_\alpha;$$
$$D_{(\beta}W_{\alpha)}| = 8 f_{\alpha\beta};$$
$$D^2 W_\alpha| = 8i\partial_{\alpha\dot\beta}\bar\lambda^{\dot\beta};$$
$$-\frac{1}{2}D^\alpha W_\alpha| = 4\mathcal{D}. \tag{4.52}$$

The $f_{\alpha\beta} = \partial_{\dot\beta\beta}A^{\dot\beta}{}_\alpha + \partial_{\alpha\dot\beta}A^{\dot\beta}{}_\beta$ here is the (symmetric) bispinor form of the usual stress tensor F_{ab}, i.e. $f_{\alpha\beta} = \frac{1}{2}(\sigma^{ab})_{\alpha\beta}F_{ab}$. In this case, the action (4.49) is reduced to

$$S = \frac{1}{4} \int d^4 x (-\frac{1}{2}f^{\alpha\beta} f_{\alpha\beta} - i\bar\lambda^{\dot\alpha}\partial_{\dot\alpha\beta}\lambda^\beta + \frac{1}{2}\mathcal{D}^2), \tag{4.53}$$

that is, the action of the free supersymmetric electrodynamics.

The key property of the action (4.53) is that it involves only higher components of the superfield V (4.24). From the formal viewpoint, it means that the lower components of this superfield, that is, $C(x)$, $\chi_\alpha(x)$, $\bar\chi^{\dot\alpha}(x)$, $M(x)$, $\bar M(x)$ can be removed via some gauge transformation. Indeed, if we consider the gauge parameter $\Lambda(z)$ whose component structure is

$$\Lambda(z) = i(\frac{1}{2}C(x) + \theta^\alpha \chi_\alpha(x) + \theta^2 M(x)), \tag{4.54}$$

with $\bar{C} = C$, we identically cancel the above-mentioned lower components of the real scalar superfield V. The gauge, in which the θ-expansion of the V (4.24) starts with $i(\bar{\theta}\sigma^a\theta)A_a(x)$, is called the Wess-Zumino gauge, and it implies that $V^n = 0$ for any $n \geq 3$. However, this gauge, although it allows to remove the nonpolynomiality of the action which seems to be very useful in the non-Abelian case, is used rarely since it breaks the superfield structure being thus very inconvenient within the superfield approach. Some attempts to conciliate this gauge with the superfield methodology have been carried out in [39].

4.2.3 Non-Abelian Gauge Theories

Now, let us generalize the gauge theory for the non-Abelian case, see e.g. [33]. The key idea consists in using the action (4.49) where the superfield strength (4.50) is promoted to a non-Abelian case. Indeed, let us suggest that it has the form

$$W_\alpha = -\bar{D}^2(e^{-gV}D_\alpha e^{gV}) \tag{4.55}$$

Here we suggest that V is a *non-Abelian*, Lie-algebra valued real scalar superfield, $V = V^A T^A$, where T^A are the Hermitian generators of some Lie group, the most popular examples of the gauge groups are $U(N)$ and $SU(N)$. It is clear that in the Abelian case this strength reduces to the expression (4.50) multiplied by the coupling g.

Let us require the theory to be invariant under the following gauge transformations:

$$e^{gV} \rightarrow e^{-ig\bar{\Lambda}}e^{gV}e^{ig\Lambda}, \tag{4.56}$$

where $\Lambda = \Lambda^A T^A$ is a Lie-algebra-valued chiral parameter of the gauge transformation, and $\bar{\Lambda}$ is an antichiral one. It is clear that the superfield strength W_α, being Lie-algebra-valued, in this case is not invariant but transformed in a covariant manner:

$$W_\alpha \rightarrow e^{-ig\Lambda}W_\alpha e^{ig\Lambda}. \tag{4.57}$$

The action of the non-Abelian gauge theory can be obtained by a straightforward promotion of the action (4.49) to the non-Abelian case (cf. [66]):

$$S = \frac{1}{64g^2} \text{tr} \int d^6z \, W^\alpha W_\alpha = -\frac{1}{16g^2} \text{tr} \int d^8z (e^{-gV} D^\alpha e^{gV}) \bar{D}^2 (e^{-gV} D_\alpha e^{gV}).$$

$$(4.58)$$

The theory described by the expression (4.58) is called the $\mathcal{N} = 1$ SYM theory. Its action is essentially non-polynomial, however, its quadratic part reproduces the Abelian expression (4.46). In principle, one can expand this action in power series in V and impose the Wess-Zumino gauge which allows to eliminate all V^3 and higher terms as well as in the Abelian case (for the discussion of the noncovariant gauges see [39]). However, the covariant gauges are much more convenient for studying this theory.

Treating the component content, one must note that the nonpolynomial gauge transformations (4.56) allow to eliminate the lower components of the superfield V. However, the non-Abelian theory (4.58) is not free but nontrivially self-coupled. We suggest that the components of the non-Abelian strength W_α are again given by the expression (4.52), with the only modification that $f_{\alpha\beta}$ is now a non-Abelian stress tensor. Therefore, the action of the SYM theory, after carrying out the procedures similar to those ones realized above is rewritten in components as

$$S = \frac{1}{4g^2} \text{tr} \int d^4x (-\frac{1}{2} f^{\alpha\beta} f_{\alpha\beta} - i \bar{\lambda}^{\dot\alpha} \nabla_{\dot\alpha\beta} \lambda^\beta + \frac{1}{2} \mathcal{D}^2), \qquad (4.59)$$

Here $\nabla_{\dot\alpha\beta} = \partial_{\dot\alpha\beta} + i A_{\dot\alpha\beta}$ is a gauge covariant space-time derivative.

The details of the supercovariant description of the $\mathcal{N} = 1$ SYM theory will be given in the Sect. 4.11, where the perturbative approach for it is discussed.

4.3 Generating Functional and Green Functions for Superfields

Now our aim consists of describing a method for calculating a generating functional and Green functions for superfields and a subsquent application of this method to calculation of superfield quantum corrections, i.e. in development of the Feynman supergraph technique. We note that during last years the activity in development of nonperturbative methods in superfield quantum theory stimulated by the paper [28] essentially increased. Nevertheless, the importance of the results obtained through the perturbative approach continues to be principal.

The generalization of the path integral method for a superfield theory turns out to be quite straightforward but a bit formal. Actually, a generating functional is defined in terms of a path integral which is well-defined only for some special cases. However, the case of the Gaussian path integral is, first, well-defined both in a standard field theory and in a superfield theory, second, sufficient for the development of the superfield perturbation technique.

Let us shortly describe the prescriptions for the introduction of Green functions in a usual field theory (see e.g. [36]). We start with the classical action $S[\phi]$ representing itself as a local space-time functional of n (super)fields ϕ^i ($i = 1, \ldots, n$) forming an (iso)vector $\vec{\phi}$. Let $\vec{\phi}_0 = \{\phi_0^i\}$ be some background. Our action $S[\phi]$ is assumed to be an analytic functional, i.e. it can be expanded into power series in a neighborhood of the $\vec{\phi}_0$:

$$S[\phi] = S[\vec{\phi}_0] + \sum_{n=2}^{\infty} \frac{1}{n!} S_{i_1 \ldots i_n} (\phi - \phi_0)^{i_n} \ldots (\phi - \phi_0)^{i_1}. \tag{4.60}$$

The term with $n = 2$ is called a quadratic (or linearized) action:

$$S_0 = \frac{1}{2} \tilde{\phi}^i S_{ij}[\vec{\phi}_0] \tilde{\phi}^j. \tag{4.61}$$

Here and further $\tilde{\phi}^i = \phi^i - \phi_0^i$. Terms with $n \geq 3$ are called interaction terms S_{int}, so, the whole action is rewritten as

$$S[\phi] = S[\vec{\phi}_0] + S_0[\tilde{\phi}^i; \vec{\phi}_0] + S_{int}[\tilde{\phi}^i; \vec{\phi}_0]. \tag{4.62}$$

The Green function G^{ij} is determined on the base of the linearized action as

$$S_{ij}[\vec{\phi}_0] G^{jk} = \delta_i^k; \quad G^{ij} S_{jk}[\vec{\phi}_0] = \delta_k^i. \tag{4.63}$$

We assume that the generating functional of Green functions is, as usual, defined in the form

$$Z[J] = N \int D\phi \exp(\frac{i}{\hbar}(S[\phi] + J\phi)). \tag{4.64}$$

Here J is an essentially classical source, and N is a normalization factor.

The propagators can be obtained on the base of the generating functional as

$$\langle \phi(z_1) \ldots \phi(z_n) \rangle = \left(\frac{1}{i} \frac{\delta}{\delta J(z_1)} \right) \ldots \left(\frac{1}{i} \frac{\delta}{\delta J(z_n)} \right) N \int D\phi \exp(\frac{i}{\hbar}(S[\phi] + J\phi)). \tag{4.65}$$

We can calculate the path integral (4.64). To do it, after relabelling $\tilde{\phi} \to \phi$ in (4.62), we expand the complete action as $S[\phi] = S_0[\phi] + S_{int}[\phi]$ with $S_0[\phi] = \frac{1}{2} \int dz \phi \Delta \phi$ is a quadratic action, and Δ be some operator. Of course, the path integration is a quite formal operation being well-defined only for the Gaussian integral and expressions derived from it. However, this is not a problem since both in the standard and in the superfield cases we need only Gaussian integrals. As usual,

$$\int D\phi \exp(\frac{i}{\hbar}(S[\phi] + J\phi)) = \int D\phi \exp(\frac{i}{\hbar}(\frac{1}{2}\phi\Delta\phi + S_{int}[\phi] + J\phi)) =$$

$$= \exp(\frac{i}{\hbar}S_{int}(\frac{\hbar}{i}\frac{\delta}{\delta J})) \int D\phi \exp(\frac{i}{\hbar}(\frac{1}{2}\phi\Delta\phi + J\phi)). \tag{4.66}$$

Since this integral is Gaussian-like, we immediately obtain

$$\int D\phi \exp(\frac{i}{\hbar}(\frac{1}{2}\phi\Delta\phi + J\phi)) = \exp(-\frac{i}{2}J(\frac{\hbar}{\Delta})J)det^{-1/2}(\frac{\Delta}{\hbar}). \tag{4.67}$$

In these expressions, the integration over the space-time is assumed where it is necessary. One concludes that all dependence of this expression on the sources is concentrated in the term $\exp(-\frac{i}{2}J(\frac{\hbar}{\Delta})J)$. Constructing of Feynman diagrams from Eqs. (4.65), (4.66), (4.67) is quite straightforward.

Let us apply this approach to a superfield theory. Our example is the Wess-Zumino model [32], and a consideration of other theories is exactly analogous. We do not address here the specifics of gauge theories in which one must introduce gauge fixing and ghosts, since afterwards all proceeding is just the same.

The action of the Wess-Zumino model with chiral sources is

$$S_J[\Phi, \bar{\Phi}; J, \bar{J}] = \int d^8z\Phi\bar{\Phi} + (\int d^6z(\frac{\lambda}{3!}\Phi^3 + \frac{m}{2}\Phi^2 + \Phi J) + h.c.). \tag{4.68}$$

As usual, conjugated terms to chiral superfields are antichiral ones. This action can be rewritten in terms of integrals over chiral and antichiral subspaces only:

$$S_J[\Phi, \bar{\Phi}; J, \bar{J}] = \int d^6z(\frac{1}{2}\Phi(-\frac{\bar{D}^2}{4})\bar{\Phi} + \frac{\lambda}{3!}\Phi^3 + \frac{m}{2}\Phi^2 + \Phi J) + h.c. \tag{4.69}$$

The generating functional is

$$Z[J, \bar{J}] = \int D\Phi D\bar{\Phi} \exp(iS_J[\Phi, \bar{\Phi}; J, \bar{J}]). \tag{4.70}$$

It is convenient to follow the approach developed in [33]. Using it, the action S_J (4.69) can be represented in matrix form

$$S_J = \frac{1}{2}\int dz_1 dz_2 \left(\Phi(z_1) \ \bar{\Phi}(z_1)\right) \begin{pmatrix} m & -\frac{1}{4}\bar{D}^2 \\ -\frac{1}{4}D^2 & m \end{pmatrix} \begin{pmatrix} \delta_+(z_1 - z_2) & 0 \\ 0 & \delta_-(z_1 - z_2) \end{pmatrix} \times$$

$$\times \begin{pmatrix} \Phi(z_2) \\ \bar{\Phi}(z_2) \end{pmatrix} + \int d^6z\Phi(z)J(z) + \int d^6\bar{z}\bar{\Phi}(\bar{z})\bar{J}(\bar{z}) + \frac{\lambda}{3!}(\int d^6z\Phi^3 + h.c.). \tag{4.71}$$

In this expression, integration in all terms is assumed with taking into account the corresponding chirality, e.g. $\int dz_1 dz_2 \Phi(z_1)m\delta_+(z_1 - z_2)\Phi(z_2) \equiv \int d^6z_1 d^6z_2 \Phi(z_1) m\delta_+(z_1 - z_2)\Phi(z_2)$, etc. We see that the operator Δ determining the quadratic part of the action (see (4.66), (4.67)) looks like

$$\Delta = \begin{pmatrix} m & -\frac{1}{4}\bar{D}^2 \\ -\frac{1}{4}D^2 & m \end{pmatrix}.$$ (4.72)

The Green function is an inverse to this operator:

$$G = \Delta^{-1} = \frac{1}{\Box - m^2} \begin{pmatrix} m & \frac{1}{4}\bar{D}^2 \\ \frac{1}{4}D^2 & m \end{pmatrix}.$$ (4.73)

In other words, the G satisfies the equation

$$\Delta G = -\begin{pmatrix} \delta_+(z_1 - z_2) & 0 \\ 0 & \delta_-(z_1 - z_2) \end{pmatrix}.$$ (4.74)

The matrix $\mathbf{1} = \begin{pmatrix} \delta_+(z_1 - z_2) & 0 \\ 0 & \delta_-(z_1 - z_2) \end{pmatrix}$ plays the role of a functional unit matrix within this description.

Thus, the generating functional can be presented as

$$Z[J, \bar{J}] = \exp(i\frac{\lambda}{3!}\left(\int d^6z \left(\frac{1}{i}\frac{\delta}{\delta J(z)}\right)^3 + h.c.\right)\det^{-1/2}\Delta \times$$

$$\times \exp\left\{-\frac{i}{2}\int dz_1 dz_2 \left(J(z_1) \ \bar{J}(z_1)\right)\frac{1}{\Box - m^2}\begin{pmatrix} m & \frac{1}{4}\bar{D}^2 \\ \frac{1}{4}D^2 & m \end{pmatrix}\times$$

$$\times \begin{pmatrix} \delta_+(z_1 - z_2) & 0 \\ 0 & \delta_-(z_1 - z_2) \end{pmatrix}\begin{pmatrix} J(z_2) \\ \bar{J}(z_2) \end{pmatrix}\right\}.$$ (4.75)

The equivalent form of this expression is

$$Z[J, \bar{J}] = \exp\left(i\frac{\lambda}{3!}\int d^6z \left(\frac{1}{i}\frac{\delta}{\delta J(z)}\right)^3 + h.c.\right)(\det^{-1/2}\Delta)\exp(\frac{i}{2}A[J, \bar{J}]) \equiv$$

$$\equiv \exp\left(i\frac{\lambda}{3!}\int d^6z \left(\frac{1}{i}\frac{\delta}{\delta J(z)}\right)^3 + h.c.\right)Z_0[J, \bar{J}],$$ (4.76)

where $Z_0[J, \bar{J}]$ is the free generating functional, and

$$A[J, \bar{J}] = -\left(\int d^6z J\frac{m}{\Box - m^2}J + 2\int d^6z J\frac{\frac{1}{4}\bar{D}^2}{\Box - m^2}\bar{J} + \int d^6\bar{z}\bar{J}\frac{m}{\Box - m^2}\bar{J}\right).$$ (4.77)

We can introduce two-point free propagators:

$$G_{++}(z_1, z_2) = \frac{1}{i^2}\frac{\delta^2 Z_0[J]}{\delta J(z_1)\delta J(z_2)}|_{J=0} = i(-\frac{1}{4})^2 \bar{D}_1^2 \bar{D}_2^2 K_{++}(z_1, z_2)$$

$$G_{+-}(z_1, z_2) = \frac{1}{i^2}\frac{\delta^2 Z_0[J]}{\delta J(z_1)\delta \bar{J}(z_2)}|_{J=0} = i(-\frac{1}{4})^2 \bar{D}_1^2 D_2^2 K_{+-}(z_1, z_2)$$

$$G_{--}(z_1, z_2) = \frac{1}{i^2}\frac{\delta^2 Z_0[J]}{\delta \bar{J}(z_1)\delta \bar{J}(z_2)}|_{J=0} = i(-\frac{1}{4})^2 D_1^2 D_2^2 K_{--}(z_1, z_2). \quad (4.78)$$

Here $K_{+-}(z_1, z_2) = K_{-+}(z_1, z_2) = -\frac{1}{\Box - m^2}\delta^8(z_1 - z_2)$, $K_{++}(z_1, z_2) = \frac{m\bar{D}^2}{4\Box(\Box - m^2)}\delta^8$ $(z_1 - z_2)$, $K_{--}(z_1, z_2) = \frac{m\bar{D}^2}{4\Box(\Box - m^2)}\delta^8(z_1 - z_2)$.

There is an alternative way to obtain the Green functions [43]. Indeed, the quadratic part of the action (4.69) can be rewritten as an integral over whole super-space:

$$S_J[\Phi, \bar{\Phi}; J, \bar{J}] = \int d^8 z \left(\Phi\bar{\Phi} + \frac{m}{2}\Phi(-\frac{D^2}{4\Box})\Phi + \frac{m}{2}\bar{\Phi}(-\frac{\bar{D}^2}{4\Box})\bar{\Phi} \right.$$
$$\left. + \Phi(-\frac{D^2}{4\Box})J + \bar{\Phi}(-\frac{\bar{D}^2}{4\Box})\bar{J} \right), \quad (4.79)$$

which has the matrix form

$$S_J[\Phi, \bar{\Phi}; J, \bar{J}] = \frac{1}{2}\int d^8 z \left(\Phi(z) \ \bar{\Phi}(z) \right) \begin{pmatrix} -m\frac{D^2}{4\Box} & 1 \\ 1 & -m\frac{\bar{D}^2}{4\Box} \end{pmatrix} \begin{pmatrix} \Phi(z) \\ \bar{\Phi}(z) \end{pmatrix} +$$
$$+ \int d^8 z \left(\Phi(z)(-\frac{D^2}{4\Box})J(z) + \bar{\Phi}(\bar{z})(-\frac{\bar{D}^2}{4})\bar{J}(\bar{z}) \right). \quad (4.80)$$

Then, since

$$\begin{pmatrix} -m\frac{D^2}{4\Box} & 1 \\ 1 & -m\frac{\bar{D}^2}{4\Box} \end{pmatrix}^{-1} = \frac{1}{\Box - m^2}\begin{pmatrix} \frac{m\bar{D}^2}{4} & \Box \\ \Box & \frac{mD^2}{4} \end{pmatrix}, \quad (4.81)$$

after the functional integration, we arrive at the expression similar to (4.76), that is,

$$Z[J, \bar{J}] = \exp\left(i\frac{\lambda}{3!}\int d^6 z \left(\frac{1}{i}\frac{\delta}{\delta J(z)}\right)^3 + h.c. \right) (\det^{-1/2}\tilde{\Delta}) \exp(-\frac{i}{2}\tilde{A}[J, \bar{J}]), \quad (4.82)$$

where $\tilde{\Delta} = \begin{pmatrix} -m\frac{D^2}{4\Box} & 1 \\ 1 & -m\frac{\bar{D}^2}{4\Box} \end{pmatrix}$, and the argument of the exponential looks like

$$\tilde{A}[J, \bar{J}] = \int d^8z \left((\tfrac{D^2}{4\Box})J(z) \; (\tfrac{\bar{D}^2}{4\Box})\bar{J}(z) \right) \begin{pmatrix} \frac{m\bar{D}^2}{4(\Box-m^2)} & \frac{\Box}{\Box-m^2} \\ \frac{\Box}{\Box-m^2} & \frac{mD^2}{4(\Box-m^2)} \end{pmatrix} \begin{pmatrix} (\tfrac{D^2}{4\Box})J(z) \\ (\tfrac{\bar{D}^2}{4\Box})\bar{J}(z) \end{pmatrix},$$

(4.83)

which can be reduced to (4.77) by straightforward transformations, i.e. the expressions (4.77) and (4.83) are equivalent. Therefore we have shown that these ways to obtain the propagators, and hence the propagators themselves, are equivalent.

We note that in a theory of a non-chiral, in particular, real scalar superfield the variational derivatives with respect to sources do not involve factors D^2, $\bar{D}^2$. These factors are caused by chirality. For example, for a theory of the real scalar superfield V, with the action $S = -\frac{1}{2} \int d^8z V \Box V$ (which emerges after an appropriate gauge fixing) the propagator is simply $G(z_1, z_2) = \frac{1}{\Box}\delta(z_1 - z_2)$.

Different vacuum expectations can be expressed in terms of the generating functional (4.75) as

$$\langle \phi(x_1) \ldots \phi(x_n) \bar{\phi}(y_1) \ldots \bar{\phi}(y_m) \rangle =$$
$$= (\frac{1}{i}\frac{\delta}{\delta J(x_1)}) \ldots (\frac{1}{i}\frac{\delta}{\delta J(x_n)})(\frac{1}{i}\frac{\delta}{\delta \bar{J}(y_1)}) \ldots (\frac{1}{i}\frac{\delta}{\delta \bar{J}(y_m)}) \times$$
$$\times \exp\left\{i\frac{\lambda}{3!} \int d^6z \left(\frac{1}{i}\frac{\delta}{\delta J(z)}\right)^3 + h.c.\right)\det{}^{-1/2}\Delta \times$$
$$\times \exp(-\frac{i}{2}\int dz_1 dz_2 \left(J(z_1)\bar{J}(z_1) \right) \frac{1}{\Box - m^2}\begin{pmatrix} m & \frac{1}{4}\bar{D}^2 \\ \frac{1}{4}D^2 & m \end{pmatrix} \times$$
$$\times \begin{pmatrix} \delta_+(z_1 - z_2) & 0 \\ 0 & \delta_-(z_1 - z_2) \end{pmatrix}\begin{pmatrix} J(z_2) \\ \bar{J}(z_2) \end{pmatrix}\right\}.$$

(4.84)

Of course, this expression contains all orders in the coupling λ. To obtain vacuum expectations up to a some order in couplings we should expand the functional operator $\exp(i\frac{\lambda}{3!}\int d^6z(\frac{1}{i}\frac{\delta}{\delta J(z)})^3 + h.c.)$ into power series in λ. As a result as usual we arrive at some Feynman diagrams. In these diagrams, there are $n + m$ external points, and the order in λ is the number of vertices. Each vertex evidently corresponds to integration over d^6z or $d^6\bar{z}$. Therefore we can introduce Feynman diagrams for superfield theories, i.e. Feynman supergraphs. Their importance consists in the fact that they allow one to preserve a manifest supersymmetric covariance at any step of calculations.

Generating functionals of arbitrary superfield models can be constructed in a whole analogy with Wess-Zumino model:

$$Z[\vec{J}] = \exp(i (S[\vec{\phi}] + \vec{\phi}\vec{J})).$$

(4.85)

Here $\vec{\phi}$ is a column matrix denoting a set of all superfields, $\vec{J}$ is a column matrix denoting a set of corresponding sources. The Green functions can be determined in analogy with (4.84). The generalization for the case of the presence of the superfields of different natures, including not only chiral and real ones but other possible superfields, does not essentially differ.

4.4 Feynman Supergraphs

Now, after we have introduced the generating functional (4.75), we can start with formulation of the superfield Feynman diagram technique, or supergraph technique. It can be introduced as follows.

One can easily read off from (4.78) that any $\langle \Phi\bar{\Phi}\rangle$-propagator corresponds to $(\Box - m^2)^{-1}$, at a chiral vertex each propagator is associated with the factor $(-\frac{1}{4}\bar{D}^2)$, and at an antichiral one—with $(-\frac{1}{4}D^2)$. However, each chiral (or antichiral) vertex involves an integration over d^6z (or $d^6\bar{z}$). Furthermore, since we deal with the delta function $\delta^8(z_1 - z_2)$, for the sake of unification it is more convenient to represent all contributions in the form of integrals over d^8z via the rule $\int d^6z(-\frac{1}{4})\bar{D}^2\mathcal{F} = \int d^8z\mathcal{F}$, with $\mathcal{F}$ be some function of superfields. As a result, if all superfields associated to a given $\int d^6z\Phi^n$-vertex are contracted into propagators, this vertex is associated with $n - 1$ $(-\frac{1}{4}\bar{D}^2)$ factors, and, similarly, any $\int d^6\bar{z}\bar{\Phi}^m$-vertex of course, in the case when all superfields are contracted into propagators—with $m - 1$ $(-\frac{1}{4}D^2)$-factors. And the vertex $\int d^8z\Phi^m\bar{\Phi}^n$, in the same case when all superfields are contracted into propagators, is associated with m factors and n $(-\frac{1}{4}D^2)$ factors. Here and further we refer to superfields contracted into propagators as to the quantum ones. The quantum chiral (antichiral) superfields will be denoted as Φ ($\bar{\Phi}$). We see that the number of D^2, $\bar{D}^2$ factors for such vertices is the number of antichiral (chiral) *quantum* superfields associated with this vertex. There is no such D^2, $\bar{D}^2$-factors arising in propagators of a non-chiral (e.g. real) superfield, since, as we argued in the Sect. 4.1, the presence of such factors is motivated by the variational derivative with respect to a chiral super-field. However, only quantum fields, i.e. those ones contracted into propagators, are associated with D^2, $\bar{D}^2$ factors at corresponding vertices. External lines do not carry such factors, and if one, two ... n chiral (antichiral) superfields associated with the vertex are external the number of $\bar{D}^2$ (D^2) factors corresponding to this vertex is less by one, two ... n than in the case when all superfields are contracted to propagators. Then, the propagator $\langle\Phi\Phi\rangle$ ($\langle\bar{\Phi}\bar{\Phi}\rangle$) corresponds to the $\frac{mD^2}{4\Box(\Box-m^2)}$ ($\frac{m\bar{D}^2}{4\Box(\Box-m^2)}$) factor, with the D^2 ($\bar{D}^2$) factor is associated with the propagator itself. Besides of this, the quantum chiral (antichiral) fields contracted to this propagator are associated to $\bar{D}^2$, D^2 by the rules defined in the beginning of this paragraph.

If we consider the $\mathcal{N} = 1$ SYM theory, its quadratic action being the sum of the action (4.49) and the simplest gauge-fixing term $S_{gf} = -\frac{1}{2}\mathrm{tr}\int d^8z V\frac{\{D^2,\bar{D}^2\}}{16}V$ corresponding to the Feynman gauge (we consider a more general gauge fixing in Sect. 4.11) looks like

$$S = -\frac{1}{2}\mathrm{tr}\int d^8z V\Box V. \tag{4.86}$$

Here tr is matrix trace (remind that the superfield V is Lie-algebra valued). There is no D factors associated with this propagator but they are associated with vertices. In a pure $\mathcal{N} = 1$ SYM theory vertices are given by

$$S_{int} = \frac{g}{16} \text{tr} \int d^8 z (\bar{D}^2 D^\alpha V)[V, D_\alpha V] + \dots \tag{4.87}$$

Here dots denote higher orders in V. The vertices of any order involve two D factors and two $\bar{D}$ factors. The D-factors in vertices involving both real and chiral (antichiral) superfields are arranged in a way described above, i.e. any vertex $\Phi \bar{\Phi} V^n$ involves one factor $(-\frac{1}{4}\bar{D}^2)$ corresponding to a Φ chiral superfield when it is contracted to the propagator and one factor $(-\frac{1}{4}D^2)$ corresponding to the $\bar{\Phi}$ antichiral superfield when it is contracted to the propagator. As a result, we can formulate Feynman rules.

The propagators look like

$$\langle \Phi(z_1)\bar{\Phi}(z_2) \rangle = -\frac{i}{\Box - m^2} \delta^8(z_1 - z_2); \tag{4.88}$$

$$\langle \Phi(z_1 0)\Phi(z_2) \rangle = \frac{im D^2}{4\Box(\Box - m^2)} \delta^8(z_1 - z_2);$$

$$\langle V(z_1)V(z_2) \rangle = \frac{i}{\Box} \delta^8(z_1 - z_2),$$

and the vertices (here Φ, $\bar{\Phi}$ are quantum superfields) correspond to

$$\int d^6 z \Phi^n \rightarrow (n-1) \text{ factors } (-\frac{1}{4})\bar{D}^2;$$

$$\int d^8 z \Phi \bar{\Phi} V^m \rightarrow \text{ one factor } (-\frac{1}{4})D^2 \text{ and one factor } (-\frac{1}{4})\bar{D}^2. \tag{4.89}$$

All derivatives associated to vertices act on the propagators. Any external chiral (antichiral) fields do not carry $\bar{D}^2$ (D^2)-factors except of those ones originated from the definition of the vertices. One will find further that the difference of overall signs of $\langle VV \rangle$ and $\langle \Phi\bar{\Phi} \rangle$ propagators plays an important role within the context of the SYM theory allowing for cancellation of some divergences. However, in principle this difference of signs takes place even in the simple non-supersymmetric scalar QED (it follows from the requirement of positivity of energy of the free scalar and gauge theories).

Within the four-dimensional Feynman supergraphs, all propagators of chiral scalar superfields will be denoted by solid lines, of gauge superfields—by wavy ones, and of ghosts—by dashed ones. The spinor supercovariant derivatives associated to vertices will be indicated explicitly.

Of course, it is more suitable to make Fourier representation for all propagators (note that Fourier transformation is carried out with respect to bosonic coordinates only) by the rule

$$\tilde{f}(k) = \int d^4 x f(x) e^{ikx}. \tag{4.90}$$

The propagators in momentum space look like

$$\langle \phi(1)\bar{\phi}(2) \rangle = \frac{i}{k^2 + m^2} \delta_{12};$$ (4.91)

$$\langle \phi(1)\phi(2) \rangle = \frac{i m D^2}{4k^2(k^2 + m^2)} \delta_{12};$$

$$\langle V(1)V(2) \rangle = -\frac{i}{k^2} \delta_{12}.$$ (4.92)

Here 1, 2 are numbers of arguments (actually we must write $V(1) \equiv V(-k, \theta_1)$ and $V(2) \equiv V(k, \theta_2)$ etc.), and δ_{12} is the purely Grassmannian delta function defined earlier. The D-factors are introduced as above. Note, however, that spinor derivatives depend after Fourier transform on a momentum of a propagator with which they are associated. The external superfields also can be represented in the form of the Fourier integral. Each propagator is parametrized by a momentum, and any vertex corresponds to an integration over $d^4\theta$, to multiplication by a corresponding coupling and a delta function over incoming momenta multiplied by $(2\pi)^4$. As usual, contribution of any supergraph includes integration over all momenta and a combinatoric factor which is defined by the same rules as in standard quantum field theory.

An essentially new feature of superfield theories consists in the presence of the D-factors. To evaluate D-algebra we can transport them from one propagator to another via integration by parts, then, we make use of the identity (4.17) in appropriate situations.

We can prove the following *non-renormalization theorem*.

The final result for the contribution of any supergraph should have the form of *one* integral over $d^4\theta$ [43].

Proof Let us consider the supergraph with L loops, V vertices and P propagators. Any vertex contains an integration over $d^4\theta$, i.e. there are V such integrations. Then, due to (4.91) any propagator carries a delta function over Grassmannian coordinates, i.e. there are P delta functions. Then, in any loop we can reduce the number of delta functions by one using Eq. (4.17), i.e. there are $P - L$ independent delta functions. As a result we can carry out $P - L$ integrations from V ones corresponding to vertices, and after D-algebra transformations we stay with $V - (P - L)$ integrations. And, due to the famous topological identity, $V - (P - L) = 1$, therefore the result contains one integration over $d^4\theta$. The theorem is proved.

This theorem means that all quantum corrections are local in θ-space. It is frequently treated naively as a proof of absence of chiral corrections which are proportional to an integral over $d^2\theta$. However, such an interpretation is wrong since any contribution in the form of an integral over the chiral subspace can be rewritten as an integral over the total superspace using the identity

$$\int d^6 z f(\Phi) = \int d^8 z \left(-\frac{D^2}{4\square}\right) f(\Phi).$$ (4.93)

Fig. 4.1 Contributions to
the two-point function of the
chiral superfield in the
Wess-Zumino model

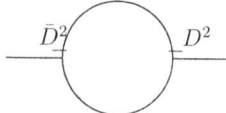

This observation was firstly made in [67], its consequences will be studied further.

Now let us study evaluation of contributions from supergraphs. The algorithm of it, after rewriting all vertices as integrals over the whole superspace, is the following one.

1. We start with one of loops. If the number of D-factors in this loop is equal to 4 we turn to the step 2. If it is more than 4, superfluous D-factors can be transported to external lines or another loops via integration by parts, and some of them are converted into internal momenta via identities $D^2 \bar{D}^2 D^2 = 16 \Box D^2$, $\{D_\alpha, \bar{D}_{\dot\beta}\} = 2i\partial_{\alpha\dot\beta}$. As a result we stay with exactly 4 D-factors. If the number of D-factors is less than 4, then the contribution from the entire supergraph is equal to zero.
2. We shrink this loop into a point using Eq. (4.17) and integrate over one of $d^4\theta$ via the delta function which is free of derivatives.
3. This procedure is repeated for next loops.
4. We integrate over internal momenta.

So, this algorithm, besides of an usual integration over internal momenta, involves additional steps, that is, D-algebra transformations. The best way to study evaluating of supergraphs consists in considering some examples.

Example 4.1 One-loop supergraph in the Wess-Zumino model (Fig. 4.1).

The contribution of this supergraph is equal to

$$
I_1 = \frac{\lambda^2}{2} \int d^4\theta_1 d^4\theta_2 \int \frac{d^4p}{(2\pi)^4} \Phi(-p, \theta_1) \bar{\Phi}(p, \theta_2) \delta_{12} \frac{\bar{D}_1^2 D_2^2}{16} \delta_{12} \times
$$

$$
\times \int \frac{d^4k}{(2\pi)^4} \frac{1}{(k^2 + m^2)((k+p)^2 + m^2)}.
\tag{4.94}
$$

The number of D-factors is just 4. D-algebra transformations are trivial: we use identity (4.17) and write $\delta_{12} \frac{\bar{D}_1^2 D_2^2}{16} \delta_{12} = \delta_{12}$. The free delta function δ_{12} allows us to integrate over $d^4\theta_2$, afterwards, we denote $\theta_1 = \theta$. As a result we get

$$
I_1 = \frac{1}{2}\lambda^2 \int d^4\theta \int \frac{d^4p}{(2\pi)^4} \Phi(-p, \theta) \bar{\Phi}(p, \theta) \int \frac{d^4k}{(2\pi)^4} \frac{1}{(k^2 + m^2)((k+p)^2 + m^2)}.
\tag{4.95}
$$

The integral over k can be calculated via dimensional regularization, the result for it is

$$
\int \frac{d^4k}{(2\pi)^4} \frac{1}{(k^2 + m^2)((k+p)^2 + m^2)} = \frac{1}{16\pi^2}(\frac{1}{\epsilon} - \int_0^1 dt \log \frac{p^2 t(1-t) + m^2}{\mu^2}).
\tag{4.96}
$$

Fig. 4.2 Two-loop vacuum
contribution in the
Wess-Zumino model

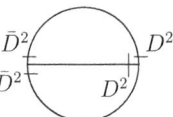

As a consequence, contribution of this supergraph takes the form

$$I_1 = \frac{1}{2}\lambda^2 \int d^4\theta \int \frac{d^4 p}{(2\pi)^4} \Phi(-p,\theta)\bar{\Phi}(p,\theta) \frac{1}{16\pi^2}(\frac{1}{\epsilon} - \int_0^1 dt \log \frac{p^2 t(1-t) + m^2}{\mu^2}).$$

(4.97)

We see that I_1 diverges logarithmically contributing to the renormalization of the kinetic term $\Phi\bar{\Phi}$.

It should be noted that in general, using of the dimensional regularization in superfield theory in higher loops possesses some peculiarities, in fact, in some cases it is ambiguous [69].

Example 4.2 Two-loop vacuum supergraph in the Wess-Zumino model (Fig. 4.2).

The contribution of this supergraph is equal to

$$I_2 = \frac{\lambda^2}{6} \int \frac{d^4 k d^4 l}{(2\pi)^8} \int d^4\theta_1 d^4\theta_2 (-\frac{\bar{D}_1^2}{4})\delta_{12} \frac{\bar{D}_1^2 D_2^2}{16}\delta_{12}(-\frac{D_2^2}{4})\delta_{12} \times$$

$$\times \frac{1}{(k^2 + m^2)(l^2 + m^2)((k+l)^2 + m^2)}.$$

(4.98)

First we do D-algebra transformations: we can write

$$(-\frac{\bar{D}_1^2}{4})\delta_{12} \frac{\bar{D}_1^2 D_2^2}{16}\delta_{12}(-\frac{D_2^2}{4})\delta_{12} = \delta_{12} \frac{\bar{D}_1^2 D_2^2}{16}\delta_{12} \frac{\bar{D}_1^2 D_2^2}{16}\delta_{12}.$$

Then we use Eq. (4.17) two times:

$$\delta_{12} \frac{\bar{D}_1^2 D_2^2}{16}\delta_{12} \frac{\bar{D}_1^2 D_2^2}{16}\delta_{12} = \delta_{12}.$$

As a result we can integrate over θ_2 using the delta function. We get

$$I_2 = \frac{\lambda^2}{6} \int \frac{d^4 k d^4 l}{(2\pi)^8} \int d^4\theta_1 \frac{1}{(k^2 + m^2)(l^2 + m^2)((k+l)^2 + m^2)}.$$

(4.99)

This integral vanishes in the standard case since it is proportional to an integral over $d^4\theta$ from a constant. However, if we suppose that the mass m is not a constant but θ-dependent superfield, this contribution will not be equal to zero. Namely this case is studied when the effective action is considered, and one uses the effective mass

Fig. 4.3 Contribution to the
two-point function in dilaton
supergravity

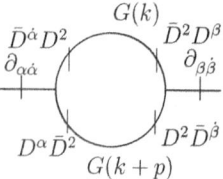

square $m^2[\Phi, \bar{\Phi}] = (m + \lambda\Phi)(m + \lambda\bar{\Phi})$, i.e. a field-dependent expression. This is
the typical prescription to calculate the Kählerian effective potential which by defi-
nition does not depend on derivatives of background superfields. This situation has
been considered in details in [73].

Example 4.3 One-loop supergraph in dilaton supergravity (Fig. 4.3).

The action of the theory looks like [63]:

$$S = \int d^8z\left[-\frac{Q^2}{16\pi^2}\bar{\sigma}\Box\sigma + \bar{D}^{\dot{\alpha}}\bar{\sigma}D^{\alpha}\sigma(\xi_1\partial_{\alpha\dot{\alpha}}(\sigma - \bar{\sigma}) + \xi_2\bar{D}_{\dot{\alpha}}\bar{\sigma}D_{\alpha}\sigma) + \frac{m^2}{2}e^{\sigma+\bar{\sigma}}\right] +$$

$$+ [\lambda \int d^6z e^{3\sigma} + h.c.]. \tag{4.100}$$

Here σ is a dimensionless chiral superfield, and $\bar{\sigma}$ is the antichiral one, so, the $\langle\sigma\bar{\sigma}\rangle$
propagators are denoted by solid lines. Other propagators are irrelevant for calculating
the divergences [63]. One of the contributions to the wave function renormalization
is given by the supergraph below. The external legs are σ and $\bar{\sigma}$.

The contribution of this supergraph is equal to

$$I_3 = \xi_1^2 \int d^4\theta_1 d^4\theta_2 \int \frac{d^4p}{(2\pi)^4}\frac{d^4k}{(2\pi)^4}(\partial_{\alpha\dot{\alpha}}\sigma(-p, \theta_1))(\partial_{\beta\dot{\beta}}\bar{\sigma}(p, \theta_2)) \times$$

$$\times \frac{D^{\alpha}\bar{D}^2 D^2 \bar{D}^{\dot{\beta}}}{16}\delta_{12}\frac{\bar{D}^{\dot{\alpha}}D^2 \bar{D}^2 D^{\beta}}{16}\delta_{12}G(k)G(k + p). \tag{4.101}$$

Here $G(k)$, $G(k + p)$ are functions of momenta whose explicit form is not essential
here, they are exactly found in [63]. As usual, within the D-algebra transformations,
the derivatives $\partial_{\alpha\dot{\alpha}}$, $\partial_{\beta\dot{\beta}}$ are not transported from external fields σ, $\bar{\sigma}$. Our aim here
is to obtain terms proportional to $\partial^m\sigma\partial^n\bar{\sigma}$ (we do not express derivatives acting
on the external fields in terms of the corresponding momenta since we will not
more manipulate with these derivatives, so, we consider $\partial_{\alpha\dot{\alpha}}\sigma$ and $\partial_{\beta\dot{\beta}}\bar{\sigma}$ as some
independent external fields. We suggest that spinor derivatives associated with one
propagator depend on the momentum k, and with another—to $k + p$.

Using commutation relations (4.13) we find that

$$\frac{D^{\alpha}\bar{D}^2 D^2 \bar{D}^{\dot{\beta}}}{16}\delta_{12}\frac{\bar{D}^{\dot{\alpha}}D^2 \bar{D}^2 D^{\beta}}{16}\delta_{12} = -\frac{4k^{\alpha\dot{\gamma}}\bar{D}_{\dot{\gamma}}D^2 \bar{D}^{\dot{\beta}}}{16}\delta_{12}\frac{4(k + p)^{\gamma\dot{\alpha}}D_{\gamma}\bar{D}^2 D^{\beta}}{16}\delta_{12}. \tag{4.102}$$

We transport all spinor supercovariant derivatives to one propagator (here the terms with spinor supercovariant derivatives moved to the external lines are omitted as the irrelevant ones since they do not contribute to the divergent part [63]). As a result we arrive at

$$16k^{\alpha\dot\gamma}(k+p)^{\gamma\dot\alpha}\delta_{12}\frac{\bar D^{\dot\beta}D^2\bar D_{\dot\gamma}D_\gamma\bar D^2 D^\beta}{256}\delta_{12}. \tag{4.103}$$

We can use (4.13) several times. At the end we get

$$16k^{\alpha\dot\gamma}(k+p)^{\gamma\dot\alpha}(k+p)_{\gamma\dot\gamma}(k+p)^{\dot\beta\delta}\delta_{12}\frac{D_\delta\bar D^2 D^\beta}{32}\delta_{12}. \tag{4.104}$$

Equations (4.13) and (4.17) allow one to write

$$\delta_{12}D_\delta\bar D^2 D^\beta\delta_{12}=-8\delta^\beta_\delta\delta_{12}.$$

We substitute this expression in (4.102). Using the identity $k^{\alpha\dot\beta}k_{\gamma\dot\beta}=\delta^\alpha_\gamma k^2$ we obtain the contribution from (4.102) in the form

$$k^{\alpha\dot\alpha}(k+p)^{\beta\dot\beta}(k+p)^2\delta_{12},$$

which after integration over θ_2 leads to the following expression for I_3:

$$I_3=-4\xi_1^2\int d^4\theta_1\int\frac{d^4p}{(2\pi)^4}\frac{d^4k}{(2\pi)^4}(\partial_{\alpha\dot\alpha}\sigma)(-p,\theta_1)(\partial_{\beta\dot\beta}\bar\sigma)(p,\theta_1)\times$$
$$\times\, k^{\alpha\dot\alpha}(k+p)^{\beta\dot\beta}(k+p)^2 G(k)G(k+p). \tag{4.105}$$

A detailed analysis carried out in [63] shows that this correction is divergent.

Calculation of corrections from supergraphs in other superfield theories is carried out in a similar manner.

We demonstrated that supergraph technique is a very efficient method for considering quantum corrections in superfield theories whereas the component study is much more complicated since one supergraph corresponds to several component diagrams (it is amusing that the exact expression for the classical action of dilaton supergravity occupies a whole page [63]). Since, as we saw in this section, the superfield theories display divergences, the next step of the development of the superfield formalism consists in introducing renormalization.

4.5 Superficial Degree of Divergence. Renormalization

We found that divergent quantum corrections arise in superfield theories as well as in usual ones. Therefore we face two problems:

(i) to classify possible divergences;
(ii) to develop a procedure of renormalization in superfield theories.

It turns out that the technique for solving these problems is quite analogous to that one used in standard field theory. The first problem can be solved on the base of the superficial degree of divergence. The natural way for solving the second one consists in introducing superfield counterterms which are quite analogous to standard ones.

First of all let us generalize the procedure to find the superficial degree of divergence (see e.g. [71]) for superfield theories. To do this, let us consider the most natural example, that is, the $\mathcal{N} = 1$ SYM theory coupled to a chiral matter with the Wess-Zumino self-interaction [33]. For all other models, the consideration is quite analogous. The action of the theory is

$$S = \int d^8 z \bar{\Phi}_i (e^{gV})^i_j \Phi^j + \left(\int d^6 z \left(\frac{1}{2} m_{ij} \Phi_i \Phi_j + \frac{\lambda_{ijk}}{3!} \Phi_i \Phi_j \Phi_k \right) + h.c. \right) - \quad (4.106)$$

$$- \mathrm{tr} \frac{1}{16g^2} \int d^8 z (e^{-gV} D^\alpha e^{gV}) \bar{D}^2 (e^{-gV} D_\alpha e^{gV}) +$$

$$+ \int d^8 z \mathrm{tr} \left(\bar{c}'c - \bar{c}c' + \frac{1}{2} g(\bar{c}' - c')[V, c + \bar{c}] + \dots \right).$$

The dots are for the higher-order couplings involving ghosts. The triple vertices in this theory are (cf. [66])

$$\frac{\lambda_{ijk}}{3!} \int d^6 z \Phi_i \Phi_j \Phi_k + h.c.; \quad g \int d^8 z \bar{\Phi}_i V^A (T^A)^i_j \Phi^j;$$

$$\frac{g}{16} \mathrm{tr} \int d^8 z (\bar{D}^2 D^\alpha V)[V, D_\alpha V]; \quad \frac{g}{2} \mathrm{tr} \int d^8 z (\bar{c}' - c')[V, c + \bar{c}]. \quad (4.107)$$

The i, j play the role of matrix indices since Φ_i is an isospinor, and the gauge superfield $V \equiv V^A T^A$ takes values in the Lie algebra as well as the ghosts do. It follows from a direct inspection of the purely gauge sector that all SYM self-interaction vertices involve exactly two chiral and two antichiral derivatives. In the previous section, we have already proved that all corrections should be proportional to one integral over $d^4 \theta$.

As usual, the superficial degree of divergence (SDD) of the given Feynman (super)graph is the order of the integral over internal momenta for the corresponding contribution, or, as is the same, is a degree of homogeneity of the (super)graph in momenta, considered after performing D-algebra transformations [33]. The only difference of the SDD in the superfield case is the additional impact from D-factors.

It is easy to see that contributions to the SDD are generated by momentum depending factors in propagators and vertices (as usual, any internal momentum k yields contribution 1), loop integrations, or, in other words, by manifest momentum dependence which is associated with propagators and loop integration, and by D-factors which are associated with propagators and vertices (note that due to identities $D^2 \bar{D}^2 D^2 = 16 \Box D^2$, $\{D_\alpha, \bar{D}_{\dot{\alpha}}\} = 2i \partial_{\alpha \dot{\alpha}}$, one chiral derivative combined with an antichiral one can be converted to one momentum, therefore any D-factor contributes to the SDD with $1/2$). If some spinor derivatives are not converted to internal momenta, the SDD from the supergraph evidently decreases.

Let us consider an arbitrary supergraph with L loops, V vertices, P propagators (C of them are $\langle \Phi\Phi \rangle$, $\langle \bar{\Phi}\bar{\Phi} \rangle$ -propagators) and E external lines (E_c of them are (anti)chiral). We denote the SDD as ω.

Any integration over internal momentum (i.e. over d^4k) contributes to ω with 4. Since the number of integrations over internal momenta is the number of loops, the total contribution from all such integrations is $4L$. Any propagator includes $\frac{1}{k^2+m^2}$ or $\frac{1}{k^2}$ (4.91), hence contribution of all propagators is equal to $-2P$. Since $\langle \Phi\Phi \rangle$, $\langle \bar{\Phi}\bar{\Phi} \rangle$-propagator contains additional $\frac{1}{k^2}$ these propagators give additional contribution $-2C$. Therefore manifest dependence of momenta gives contribution to ω equal to $4L - 2P - 2C$.

Now let us consider contribution of D-factors to the SDD. Each vertex (both pure gauge one and that one containing chiral superfields) without external chiral (antichiral) lines contains four D-factors since any superfield Φ (contracted to propagator) corresponds to $\bar{D}^2$, and $\bar{\Phi}$—to D^2. Therefore each vertex yields the contribution 2. However, external chiral (antichiral) lines do not carry D-factors. As a result, any external Φ, $\bar{\Phi}$ line decreases ω by 1. Each $\langle \Phi\Phi \rangle$, $\langle \bar{\Phi}\bar{\Phi} \rangle$-propagator contains a factor $\bar{D}^2$ (D^2) with contribution 1. Then, due to the identity (4.17), contracting any loop into a point decreases the number of D-factors which can be converted to internal momenta by 4, consequently, ω—by 2. As a result the total contribution of D-factors to ω is equal to $2V - E_c - 2L + C$ (remind that each D-factor contributes to ω with $1/2$).

Therefore the SDD is equal to

$$\omega = 4L - 2P - 2C + 2V - E_c - 2L + C = 2L - 2P + 2V - C - E_c. \tag{4.108}$$

Using the well known topological identity $L + V - P = 1$ we have

$$\omega = 2 - C - E_c. \tag{4.109}$$

Really, the SDD can be lower than (4.109) if some of D-factors are transported to external lines and do not generate internal momenta. If N_D D-factors are moved to external lines the ω is equal to

$$\omega = 2 - C - E_c - \frac{1}{2}N_D. \tag{4.110}$$

This is the final expression for the SDD. As usual, at $\omega \geq 0$ a supergraph diverges, and at $\omega < 0$—converges. We note that:

1. $\omega \leq 2$ hence the SDD is restricted from above.
2. As the number of external lines grows (here for the sake of simplicity we consider only external (anti)chiral legs while the presence of external gauge legs will be treated in the Sect. 4.11), ω decreases. Therefore the number of divergent structures is essentially limited—it is finite (really, there can be no more than

two external chiral legs and no more than two $\langle\Phi\Phi\rangle$, $\langle\bar{\Phi}\bar{\Phi}\rangle$ propagators in a superficially divergent supergraph). And since the number of divergent structures is finite, the theory is renormalizable. Hence we have just shown that the $\mathcal{N} = 1$ SYM theory coupled to the chiral matter with the Wess-Zumino self-interaction, described by action (4.106) is renormalizable. This is quite natural since the mass dimensions of all couplings in this theory are equal to zero.

However, non-renormalizable superfield theories also exist. The example is the general chiral superfield model [62].

The action of the model is

$$
S = \int d^8z K(\Phi, \bar{\Phi}) + (\int d^6z W(\Phi) + h.c.) =
$$

$$
= \int d^8z(\Phi\bar{\Phi} + \sum_{m+n\geq 3}^{\infty} \frac{1}{m!n!} K_{mn}\Phi^n\bar{\Phi}^m) + [\int d^6z(\frac{m}{2}\Phi^2 + \sum_{l=3}^{\infty} \frac{W_l}{l!}\Phi^l) + h.c.].
$$

Here K_{mn}, W_l are constants with nontrivial (actually, negative) mass dimensions.

Propagators in this theory are just (4.91), their contribution, together with loop integrations, to the SDD is equal to $4L - 2P - 2C$ as above. However, the contribution from D-factors differs from the Wess-Zumino case. Any vertex $K_{nm}\Phi^n\bar{\Phi}^m$ corresponds to n $\bar{D}^2$-factors and m D^2-factors. The total contribution to ω from all such vertices is the sum of n's and m's taken over all vertices defined as integrals over the total superspace (this is denoted by the subscript V_t), i.e. $\sum_{V_t}(n_{V_t} + m_{V_t})$, where n_{V_t}, m_{V_t} are just numbers n, m for any given vertex defined in the whole superspace. Any vertex $W_l\Phi^l$ contains an integral over d^6z and effectively corresponds to $(l - 1)$ $\bar{D}^2$-factors (for the analogous antichiral vertex, to $(l - 1)$ D^2 factors). The total contribution from such vertices is the sum over all purely chiral or antichiral vertices (this is denoted by the subscript V_c), i.e. $\sum_{V_c}(l_{V_c} - 1)$, where l_{V_c}'s are numbers l for any given chiral vertex. As in the Wess-Zumino model, external chiral (antichiral) lines decrease the number of D^2 ($\bar{D}^2$)-factors by $2E_c$, where E_c is a number of external lines, each $\langle\Phi\Phi\rangle$, $\langle\bar{\Phi}\bar{\Phi}\rangle$-propagator carries one $\bar{D}^2$ (D^2)-factor. Contracting of each loop to a point decreases the number of D-factors by 4. Thus, the total number of D-factors is

$$
2\sum_{V_t}(n_{V_t} + m_{V_t}) + 2\sum_{V_c}(l_{V_c} - 1) - 2E_c - 4L + 2C. \tag{4.111}
$$

Contribution to the SDD from D-factors is their number divided by two. Therefore the total SDD in this theory is equal to

$$
\omega = 4L - 2P - 2C + \frac{1}{2}(2\sum_{V_t}(n_{V_t} + m_{V_t}) + 2\sum_{V_c}(l_{V_c} - 1) - 2E_c - 4L + 2C) =
$$

$$
= 2 - 2V - C - 2E_c + [\sum_{V_t}(n_{V_c} + m_{V_c}) + \sum_{V_c}(l_{V_c} - 1)]. \tag{4.112}
$$

Here we used $2L - 2P = 2 - 2V$. However, any vertex gives contribution -2 to the term $-2V$ and $l_{V_c} - 1$ or $n_{V_t} + m_{V_t}$ to terms of ω involving summations. It is evident that either $l_{V_c} - 1$ or $n_{V_t} + m_{V_t}$ is not less than 2 since either $l_{V_c} \geq 3$ or $n_{V_t} + m_{V_t} \geq 2$. Hence in a general case $\sum_{V_t}(n_{V_t} + m_{V_t}) + \sum_{V_c}(l_{V_c} - 1) - 2V \geq 0$, thus the number of divergent structures is not restricted, and the theory is non-renormalizable. This is quite natural since the constants K_{ij} (for $i + j > 2$) and W_l (for $l \geq 4$) have negative mass dimensions.

The next problem consists in introducing a regularization scheme. The most natural way to do it in field theories, including supersymmetric ones, is the dimensional regularization. It can be implemented as usual: any integral

$$\int \frac{d^4k}{(2\pi)^4} \frac{1}{(k^2 + m^2)^N}$$

is replaced by its extension to $(4 + \epsilon)$-dimensional space-time.

$$\mu^{-\epsilon} \int \frac{d^{4+\epsilon}k}{(2\pi)^{4+\epsilon}} \frac{1}{(k^2 + m^2)^N},$$

so, all divergences are described by poles in ϵ, as in the usual field theory (no more than $\frac{1}{\epsilon^L}$ for a L-loop correction).

However, there are some peculiarities. First of all, within the component description any supersymmetric action includes spinors and hence γ-matrices which are well defined if and only if the dimension of the space-time is integer. Therefore we must use some modification of the dimensional regularization called dimensional reduction. According to it, all objects which are well-defined only for specific dimensions, for example integer ones (such as spinors and Dirac γ matrices) are evaluated at these dimensions (in our case—at the dimension equal to 4), and integrals over momenta—at arbitrary dimensions. At the same time, the dimensional reduction leads to some difficulties in calculation of higher loop corrections since many supergraphs involve contractions of essentially four-dimensional objects, such as Levi-Civita tensor ϵ^{abcd}, with d-dimensional objects, and such contractions need additional definitions. As a result the ambiguities frequently arise. However, such a situation is observed only beyond two loops. The detailed discussion of different problems related with applying the dimensional regularization in supersymmetric field theories is presented in [69].

The technique for renormalization in superfield theories is quite analogous to that one in common QFT. It is carried out via introducing the corresponding counterterms.

Example. Let us consider the one-loop contribution to the kinetic term in the Wess-Zumino model. The corresponding supergraph is given by Fig. 4.1 (see above), its contribution (4.97) yields

$$I_1 = \frac{1}{2}\lambda^2 \int d^4\theta \int \frac{d^4p}{(2\pi)^4} \Phi(-p,\theta)\bar{\Phi}(p,\theta) \frac{1}{16\pi^2}(\frac{1}{\epsilon} - \int_0^1 dt \log \frac{p^2 t(1-t) + m^2}{\mu^2}).$$

$$(4.113)$$

We see that this divergence has the form of pole part proportional to $\frac{1}{\epsilon}$. To cancel it we must add to the initial kinetic term

$$S = \int d^8z \Phi(x,\theta)\bar{\Phi}(x,\theta),$$

$$(4.114)$$

the counterterm

$$\Delta S_{countr} = -\frac{\lambda^2}{32\pi^2\epsilon} \int d^8z \Phi(z)\bar{\Phi}(z)$$

$$(4.115)$$

which corresponds to the replacement of $\int d^8z \Phi\bar{\Phi}$ in the classical action by $Z \int d^8z \Phi(z)\bar{\Phi}(z)$ where

$$Z = 1 - \frac{\lambda^2}{32\pi^2\epsilon}$$

$$(4.116)$$

is the wave function renormalization.

The essential advantage of supersymmetric theories is the fact that the number of counterterms in these theories is less than in their non-supersymmetric analogues. For example, Wess-Zumino model is a supersymmetric generalization of ϕ^4-theory, but it involves only one renormalization constant corresponding to the renormalization of kinetic term and no renormalization of couplings. The conclusion about absence of divergent correction to the coupling λ (or as is the same—to the chiral potential) is also called non-renormalization theorem since it is treated as a consequence of the non-renormalization theorem discussed in the Sect. 4.4, following which, all loop corrections are local in superspace involving only one integral over $d^4\theta$. However, actually, the existence of the *finite* corrections to the chiral (holomorphic) potential is not forbidden, for example, they arise in the massless Wess-Zumino model [25, 26, 72, 73]. In the Sect. 4.8 we discuss such corrections.

There are also some interesting properties of renormalization in superfield theories.

First, all tadpoles in the Wess-Zumino model (see Fig. 4.4) vanish. Indeed, such a supergraph has a contribution proportional to $D^2\delta_{11} = \delta_{12}D^2\delta_{12} = 0$. The similar situation can occur in other superfield models involving the Wess-Zumino model as an ingredient. However, in theories including vertices proportional to an integral over the whole superspace (e.g. dilaton supergravity) tadpole contributions are not equal to zero [63].

Second, all contributions from vacuum supergraphs are proportional to $\int d^4\theta c$ (with c is a constant) and also vanish. However, this statement is not true for background dependent propagators. The methodology of background dependent propa-

Fig. 4.4 Tadpole
contribution in the
Wess-Zumino model

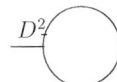

gators naturally arises within the formalism of the effective action whose key features
are the same in usual field theories (see Chap. 2) and in supersymmetric ones, hence,
the concepts developed in the Chap. 2 can be straightforwardly applied for the superfield theories. In the next section we carry out this application.

4.6 Effective Action in Superfield Theories. Superfield Proper-Time Technique

Let us now discuss the problem of the effective action in superfield theories. In
the Chap. 2, we have shown that the effective action $\Gamma[\Phi]$ depending on the
background (super)field Φ (actually, in the general case the Φ denotes a set of
all background (super)fields, whose indices are suppressed) can be presented as
$\Gamma[\Phi] = S[\Phi] + \bar{\Gamma}[\Phi]$, where $S[\Phi]$ is a classical action of the theory, and $\bar{\Gamma}[\Phi]$ is
the complete quantum contribution to the effective action which can be determined
from the expression

$$e^{\frac{i}{\hbar}\bar{\Gamma}[\Phi]} = \int D\phi \, e^{\frac{i}{2}S''[\Phi]\phi^2} \left(1 + \frac{i\sqrt{\hbar}}{3!}S^{(3)}[\Phi]\phi^3 + \frac{i\hbar}{4!}S^{(4)}[\Phi]\phi^4 + \right.$$
$$\left. + \frac{1}{2}\left(\frac{i\sqrt{\hbar}}{3!}\right)^2 (S^{(3)}[\Phi]\phi^3)^2 + \ldots \right). \tag{4.117}$$

Expanding the effective action in power series in $\hbar$ as $\bar{\Gamma}[\Phi] = \sum_{L=1}^{\infty} \hbar^L \Gamma_L[\Phi]$, one
can express the one-loop contribution to $\Gamma[\Phi]$ in terms of the trace of the logarithm
of the background dependent operator $S''[\Phi]$ as

$$\Gamma^{(1)} = \frac{i}{2}\text{Tr}\ln\Delta[\Phi], \tag{4.118}$$

where $\Delta[\Phi] \equiv S''[\Phi]$ is an operator characterizing the quadratic action of the quantum fields. This is the famous expression of the one-loop effective action in terms of
the trace of the logarithm. Studying of the one-loop correction is a starting point for
any discussions of the effective action.

Now, it is instructive to discuss the following question: how the definition of the
one-loop correction in an effective action in terms of the trace of the logarithm is
related to the expression of the same correction in terms of (super)graphs?

To clarify this relation we give an example. The one-loop effective action in the
Wess-Zumino model is given by the following functional trace [27, 33]:

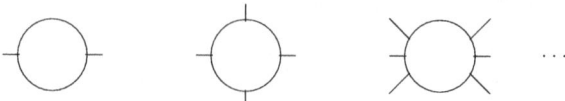

Fig. 4.5 Feynman supergraphs arising from the expansion of the trace of the logarithm

$$\Gamma^{(1)} = \frac{i}{2} \text{Tr} \log(\Box - \frac{1}{4} \Psi \bar{D}^2 - \frac{1}{4} \bar{\Psi} D^2). \tag{4.119}$$

Here $\Psi = m + \lambda \Phi$ is the background chiral superfield. It is clear that the operator Δ in this case is $\Delta = \Box - \frac{1}{4} \Psi \bar{D}^2 - \frac{1}{4} \bar{\Psi} D^2$. The $\Gamma^{(1)}$ can be rewritten as

$$\Gamma^{(1)} = \frac{i}{2} \text{Tr} \log[\Box(1 - \frac{1}{4\Box}(\Psi \bar{D}^2 + \bar{\Psi} D^2))]. \tag{4.120}$$

Expansion of the logarithm into power series leads to

$$\Gamma^{(1)} = -\frac{i}{2} \text{Tr} \sum_{n=1}^{\infty} \frac{1}{n} [\frac{1}{4\Box}(\Psi \bar{D}^2 + \bar{\Psi} D^2)]^n. \tag{4.121}$$

This expression exactly reproduces the total contribution for the sum of the supergraphs depicted at Fig. 4.5.

Here, external lines are for alternating Ψ and $\bar{\Psi}$ fields, with the $-\frac{D^2}{4}$ and $-\frac{\bar{D}^2}{4}$ factors are associated with the vertices as usual, and internal ones are for the free propagator of the chiral superfield. At the same time, if we consider a theory of a real scalar superfield u in the external chiral superfield Ψ with action

$$S = \frac{1}{2} \int d^8z u (\Box - \frac{1}{4} \Psi \bar{D}^2 - \frac{1}{4} \bar{\Psi} D^2) u, \tag{4.122}$$

and treat the $\int d^8z u(-\frac{1}{4} \Psi \bar{D}^2) u$ and the conjugated term as interaction vertices, we arrive just at these supergraphs, and one-loop effective action for this theory is again given by (4.119).

We can see that the expression of the one-loop effective action in the form of the trace of the logarithm of some operator allows to use some special technique which is equivalent to supergraph approach, but more convenient in many cases. This technique is called the proper-time method. Its essence is as follows.

Let us start with the assumption that the quadratic action of a quantum (super)field ϕ on a classical background Φ has the form $\frac{1}{2} \int dx \phi \Delta[\Phi] \phi$, where $\int dx$ here denotes the integral over all (super)space, and the $\Delta[\Phi]$ is an operator which in typical cases looks like $\Delta = \Box + \dots$, where dots denote background dependent terms. As we argued above, the one-loop effective action in this theory can be presented as $\Gamma^{(1)} = \frac{i}{2} Tr \int_0^\infty \frac{ds}{s} e^{is\Delta}$. Therefore we face the problem of calculating the operator $e^{is\Delta}$. It is known [36] that the best way to find this operator in the case of usual field theories is as follows. We introduce the function $U(x, x'|s) = e^{is\Delta} \delta^4(x - x')$

called the Schwinger kernel. Of course, the U depends on background superfields. It follows from this definition that U satisfies the equation:

$$i\frac{\partial U}{\partial s} = -U\Delta. \tag{4.123}$$

The Δ is supposed to have a form of a power series in derivatives, and U by the definition satisfies the initial condition

$$U(x, x')|_{s=0} = \delta^4(x - x').$$

In a usual case (especially, in the gravity theories) the U is represented in the form of infinite power series in parameter s, called the proper time, as [36]

$$U = -\frac{i}{(4\pi s)^2} \exp(\frac{i}{4s}(x - x')^2) \sum_{n=0}^{\infty} a_n (is)^n. \tag{4.124}$$

The (ultraviolet) divergences correspond to *lower* orders of this expansion, one must note that the ultraviolet limit corresponds to $s \to 0$, infrared one—to $s \to \infty$. Coefficients a_n depend on background fields and their derivatives. We note that if background fields are put to zero, we arrive at

$$U^{(0)}(x, x'; s) = e^{is\square}\delta^4(x - x') = -\frac{i}{(4\pi s)^2} \exp(\frac{i}{4s}(x - x')^2), \tag{4.125}$$

and the $U^{(0)}(x, x'; s)$ satisfies the condition

$$i\int_0^\infty ds\, U^{(0)}(x, x'; s) = \frac{1}{\square}\delta^4(x - x'). \tag{4.126}$$

The approach in the case of superfield theories is quite analogous. However, in the superfield case using of the proper time approach is characterized by an essential advantage. Indeed, in this case it is more convenient to expand Schwinger kernel $U(x, x'; s)$ not in an infinite power series in s but in a power series in spinor supercovariant derivatives which is *finite* due to anticommutation properties of spinor derivatives (we already know that the similar situation takes place in three-dimensional superfield theories).

Actually, in most cases the operator Δ in superfield theories looks like

$$\Delta = \square + A_{10}^\alpha D_\alpha + A_{01\dot\alpha}\bar{D}^{\dot\alpha} + \ldots \equiv \square + \tilde{\Delta}, \tag{4.127}$$

with $\tilde{\Delta}$ is a some background dependent operator. In typical cases it contains only even orders in spinor derivatives, here we consider just this case, and A_{nm}'s are background dependent coefficients. We introduce the superfield Schwinger kernel (cf. [27, 33]):

$$U(z, z'; s) = \exp(is\Delta)\delta^8(z - z') \equiv \exp(is\tilde{\Delta})\exp(is\Box)\delta^8(z - z'). \quad (4.128)$$

The last identity is valid for studying of contributions which do not depend on space-time derivatives of superfields, i.e. for contributions to the effective potential. Then, we suppose a natural initial condition

$$U(z, z'; s)|_{s=0} = \delta^8(z - z').$$

It is clear that $\exp(is\Box)\delta^8(z - z') = \delta^4(\theta - \theta')U^{(0)}(x, x'; s)$ where $U^{(0)}(x, x'; s)$ is given by (4.125). Hence

$$U(z, z'; s) = \exp(is\tilde{\Delta})U^{(0)}(x, x'; s)\delta^4(\theta - \theta'). \quad (4.129)$$

Therefore we face the problem of calculating the operator $\Omega = \exp(is\tilde{\Delta})$. The Ω satisfies the equation

$$i\frac{\partial\Omega}{\partial s} = -\Omega\tilde{\Delta}. \quad (4.130)$$

It is easy to see that $\Omega|_{s=0} = 1$. We expand Ω into a finite power series in spinor supercovariant derivatives (its finiteness is based on the anticommutation relations of the derivatives, cf. [27]; the linear terms in D_α, $\bar{D}_{\dot\alpha}$ in typical cases are unnecessary):

$$\Omega = 1 + \frac{1}{16}A(s)D^2\bar{D}^2 + \frac{1}{16}\tilde{A}(s)\bar{D}^2D^2 + \frac{1}{8}B^\alpha(s)D_\alpha\bar{D}^2 + \frac{1}{8}\tilde{B}_{\dot\alpha}(s)\bar{D}^{\dot\alpha}D^2 +$$
$$+ \frac{1}{4}C(s)D^2 + \frac{1}{4}\tilde{C}(s)\bar{D}^2, \quad (4.131)$$

and substitute (4.131) into the equation (4.130). As a result we obtain some power series in spinor derivatives in the r.h.s. of (4.130). Comparing coefficients at analogous combinations of the spinor derivatives in the r.h.s. and in the l.h.s. of this equation, we get [27]:

$$\frac{1}{16}\dot{A} = \Omega\tilde{\Delta}|_{D^2\bar{D}^2};$$
$$\frac{1}{8}\dot{B}^\alpha = \Omega\tilde{\Delta}|_{D_\alpha\bar{D}^2};$$
$$\frac{1}{4}\dot{C} = \Omega\tilde{\Delta}|_{D^2} \quad (4.132)$$

and analogous equations for $\tilde{A}$, $\tilde{B}_{\dot\alpha}$, $\tilde{C}$. Here the dot denotes $\frac{1}{i}\frac{\partial}{\partial s}$, and $|_{D^2}$ etc. denotes the coefficient at D^2 etc. in the expansion of $\Omega\tilde{\Delta}$. As a result we have a system of first-order differential equations for coefficients determining the structure of the operator Ω. Since $\Omega|_{s=0} = 1$, we have natural initial conditions

$$A|_{s=0} = \tilde{A}|_{s=0} = B^\alpha|_{s=0} = \tilde{B}_{\dot\alpha}|_{s=0} = C|_{s=0} = \tilde{C}|_{s=0} = 0. \qquad (4.133)$$

The system (4.132) with initial conditions (4.133) can be solved in a manner similar to a common system of differential equations. However, one must notice that this solution can be exactly found only in special cases, for example, for the dependence of the heat kernel only on background superfields but not on their derivatives, or for its dependence on chiral background superfields only.

Then, the $\Omega(s)$, sometimes also referred as a heat kernel, can be used for the calculation of the Green function as

$$G(z_1, z_2) = i \int_0^\infty ds\, \Omega U^{(0)}(x, x'; s)\delta^4(\theta - \theta') \qquad (4.134)$$

(note that the Ω is a differential operator in the superspace), and for the calculation of the one-loop effective action as

$$\Gamma^{(1)} = \frac{i}{2} \int_0^\infty \frac{ds}{s} \int d^8z\, d^8z'\, \delta^8(z - z')\Omega U^{(0)}(x, x'; s)\delta^4(\theta - \theta'). \qquad (4.135)$$

As usual, $\int d^8z = \int d^4x\, d^4\theta$, we also use the definition (4.125). Then, it is known that $\delta^4(\theta - \theta')D^2\bar{D}^2\delta^4(\theta - \theta') = 16\delta^4(\theta - \theta')$, and all products of less number of spinor derivatives give a zero trace. Hence only coefficients of (4.131) giving non-zero contribution to one-loop effective action are A and $\tilde{A}$. And the one-loop effective action looks like

$$\Gamma^{(1)} = \frac{i}{2} \int_0^\infty \frac{ds}{s} \int d^4x\, d^4\theta (A(s) + \tilde{A}(s))U^{(0)}(x, x'; s)|_{x=x'}. \qquad (4.136)$$

Thus, we presented a technique proposed in [27] for calculating background dependent propagators and one-loop effective action. Application of this technique will be further considered for examples of several theories. There exists an essential modification of this method for supersymmetric gauge theories [74]. We discuss it further.

4.7   Problem of Superfield Effective Potential

The effective potential in a standard quantum field theory is defined as the effective Lagrangian evaluated at constant values of scalar fields, and other fields are put to zero. The effective potential is used for studying of spontaneous symmetry breaking and vacuum stability [34], as well as for many other issues.

First, let us shortly describe the effective potential in a usual quantum field theory. The effective action can be presented as a derivative expansion:

$$\Gamma[\phi] = \int d^4x(-V_{eff}(\phi) + \frac{1}{2}Z(\phi)\partial_m\phi\partial^m\phi + \ldots),\qquad(4.137)$$

where $Z(\phi)$ is some function of ϕ, and $V_{eff}(\phi)$ is the effective potential. Therefore, for slowly varying fields one has

$$\Gamma[\phi] = -\int d^4x\, V_{eff}(\phi),$$

so, effective potential is a low-energy leading term. It can be represented in the form of the loop expansion

$$V_{eff}(\phi) = V(\phi) + \sum_{n=1}^{\infty} \hbar^n V^{(n)}(\phi).\qquad(4.138)$$

For example, let us consider the theory with the action

$$S = \int d^4x(-\frac{1}{2}\phi\Box\phi - V(\phi)).\qquad(4.139)$$

After background-quantum splitting $\phi \to \Phi + \chi$ where Φ is the background super-field and χ is the quantum one, we find the quadratic action of quantum superfields

$$S_2 = -\frac{1}{2}\int d^4x\chi(\Box + V''(\Phi))\chi,\qquad(4.140)$$

which leads to the one-loop effective action $\Gamma^{(1)}[\Phi]$ of the form

$$\Gamma^{(1)}[\Phi] = \frac{i}{2}\text{Tr}\log(\Box + V''(\Phi)).\qquad(4.141)$$

Following the previous studies, we can express this trace of the logarithm in the form of diagrams depicted at Fig. 4.6, where external lines are $V''(\Phi)$. Internal lines correspond to massless scalar propagators.

The complete effective potential is presented by a sum of contributions from these Feynman diagrams:

Fig. 4.6 Feynman supergraphs arising from the expansion of Eq. (4.141)

  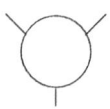 $\cdots$

$$U^{(1)} = \sum_{n=1}^{\infty} \frac{1}{2n} \int \frac{d^4k}{(2\pi)^4} \left(\frac{V''(\Phi)}{k^2}\right)^n = -\int \frac{d^4k}{(2\pi)^4} \log\left(1 - \frac{V''(\Phi)}{k^2}\right). \quad (4.142)$$

Integrating over d^4k and subtracting the divergence (cf. [34]), we arrive at

$$U^{(1)} = -\frac{1}{32\pi^2}(V''(\Phi))^2\left(\log \frac{V''(\Phi)}{\mu^2} + C\right), \quad (4.143)$$

where C is a constant which can be fixed by imposing of some renormalization conditions (see [34] for details in the case of $\lambda\phi^4$ theory). The same result can be also obtained via the proper-time method.

Now we turn to a superfield case. Let $\Gamma[\Phi, \bar{\Phi}]$ be the (renormalized) effective action for a theory of chiral and antichiral superfields. We can represent it as [27]

$$\Gamma[\bar{\Phi}, \Phi] = \int d^8z \mathcal{L}_{eff}(\Phi, D_A\Phi, D_A D_B\Phi; \bar{\Phi}, D_A\bar{\Phi}, D_A D_B\bar{\Phi}) +$$

$$+ \left(\int d^6z W_{eff}(\Phi) + h.c.\right) + \ldots \quad (4.144)$$

Here $D_A\Phi, D_A D_B\Phi, \ldots$ are spinor supercovariant derivatives of superfields $\Phi, \bar{\Phi}$. The term $\mathcal{L}_{eff}$ is called the general effective Lagrangian, and W_{eff} is called the chiral effective Lagrangian. Both these effective Lagrangians can be expanded into power series in supercovariant derivatives of background superfields. The dots in this expression denote terms depending on space-time derivatives of $\Phi, \bar{\Phi}$. Further, the structure of the effective action (4.144) will be considered as a standard one for the superfield theories. We note that since the chiral effective Lagrangian by definition depends only on Φ but not on $\bar{D}^2\Phi$, all terms of the form

$$\int d^6z \Phi^n (\bar{D}^2\bar{\Phi})^m,$$

using the relation $\int d^6z(-\frac{\bar{D}^2}{4}) = \int d^8z$, can be rewritten as

$$\int d^8z \Phi^n \bar{\Phi}(\bar{D}^2\bar{\Phi})^{m-1},$$

i.e. in the form corresponding to the general effective Lagrangian. Therefore here and further we consider all expressions which are formally chiral but involve $(\bar{D}^2\bar{\Phi})^m$ as contributions to the general effective Lagrangian.

We note that all chiral contributions can be also represented as an integral over the whole superspace (this observation has been made for the first time in [67]):

$$\int d^6z G(\Phi) = \int d^8z\left(-\frac{D^2}{4\Box}\right)G(\Phi). \quad (4.145)$$

Further, to recover the usual effective potential within the component approach we must put scalar component fields to be constant, and spinor ones—to zero, e.g. in the Wess-Zumino model we write

$$A = \text{const}, \quad F = \text{const}, \quad \psi_\alpha = 0.$$

However, this condition is not supersymmetric, therefore instead of it, we use condition for the superfield to be *constant in the space-time*:

$$\partial_m \Phi = 0. \tag{4.146}$$

Since ∂_m commutes with all generators of supersymmetry, this condition is supersymmetric.

The effective potential is introduced as

$$V_{eff} = \left\{ -\int d^4\theta \mathcal{L}_{eff} - \left(\int d^2\theta W_{eff} + h.c.\right) \right\}\Big|_{\partial_a \Phi = \partial_a \bar{\Phi} = 0}. \tag{4.147}$$

The minus sign is put by convention. We can introduce a general effective potential $\mathcal{L}_{eff}|_{\partial_a \Phi = \partial_a \bar{\Phi} = 0}$ and a chiral (or holomorphic, as is the same) effective potential $W_{eff}|_{\partial_a \Phi = 0}$. It is easy to see that the general effective potential can be expressed as

$$\mathcal{L}_{eff} = \mathbf{K}(\Phi, \bar{\Phi}) + \mathbf{F}(D_\alpha \Phi, \bar{D}_{\dot{\alpha}} \bar{\Phi}, D^2 \Phi, \bar{D}^2 \bar{\Phi}; \Phi, \bar{\Phi}) \tag{4.148}$$

with $\mathbf{F}|_{D_\alpha \Phi, \bar{D}_{\dot{\alpha}} \bar{\Phi}, D^2 \Phi, \bar{D}^2 \bar{\Phi} = 0} = 0$. The $\mathbf{K}(\Phi, \bar{\Phi})$ is called the Kählerian effective potential, and the $\mathbf{F}(D_\alpha \Phi, \bar{D}_{\dot{\alpha}} \bar{\Phi}, D^2 \Phi, \bar{D}^2 \bar{\Phi}; \Phi, \bar{\Phi})$ is called the auxiliary fields' effective potential, it is at least of third order in auxiliary fields of Φ and $\bar{\Phi}$. These objects can be represented in the form of the loop expansion:

$$\mathbf{K}(\Phi, \bar{\Phi}) = K_0(\Phi, \bar{\Phi}) + \sum_{L=1}^{\infty} \hbar^L K^{(L)}(\Phi, \bar{\Phi}), \tag{4.149}$$

$$\mathbf{F} = \sum_{L=1}^{\infty} \hbar^L F^{(L)}, \tag{4.150}$$

(the term corresponding to tree level, $L = 0$, in the expression for $\mathbf{F}$ is absent for theories which do not include derivative depending terms in the classical action, such as the Wess-Zumino model), and

$$W_{eff}(\Phi) = W(\Phi) + \sum_{L=1}^{\infty} \hbar^L W^{(L)}(\Phi). \tag{4.151}$$

Here $K^{(L)}$, $F^{(L)}$, $W^{(L)}$ are quantum corrections. For the Wess-Zumino model, $W^{(1)} = 0$, however, in some quantum theories (e.g. in $\mathcal{N} = 1$ SYM theory with chiral matter) a one-loop contribution to chiral effective potential exists [26].

The structure of the effective potential presented by the Eqs. (4.147)–(4.151) is generic describing all theories of the chiral superfields (remind that within the phenomenological context, the matter is associated with chiral superfields since namely they involve the usual scalar fields as components) including the noncommutative ones. However, we note that the effective potential in theories including gauge superfields must depend on these superfields in a special way. Indeed, the effective action in such theories should be expressed in terms of some gauge invariant constructions. For example, within the background field method, the gauge superfield is incorporated either into covariantly chiral superfields or into supercovariant derivatives and gauge invariant superfield strengths [33, 66].

Let us give a few remarks about the method of calculating the effective potential. The best way for it, of course, is based on the using of background dependent propagators which are expressed in terms of usual propagators and background superfields. Background dependent propagators can be exactly found in certain cases. To calculate Kählerian effective potential and auxiliary fields' effective potential one can straightforwardly omit all space-time derivatives of background superfields, moreover, to study Kählerian effective potential one can omit *all* supercovariant derivatives and treat background superfields as constants until the final integration. The calculation of the chiral effective potential, however, is characterized by some peculiarities. The best example to illustrate it is the Wess-Zumino model—the simplest superfield theory. We will consider it in the next section.

4.8 The Wess-Zumino Model and a Problem of the Chiral Effective Potential

Now we turn our attention to considering the superfield effective potential in the Wess-Zumino model. Here we follow the papers [27, 72, 73] and the book [33].

The superfield action of the Wess-Zumino model is given by (4.68). Following, as usual, the loop expansion approach, we carry out background-quantum splitting by the rule

$$\begin{aligned} \Phi &\to \Phi + \sqrt{\hbar}\phi; \\ \bar{\Phi} &\to \bar{\Phi} + \sqrt{\hbar}\bar{\phi}, \end{aligned} \qquad (4.152)$$

where again Φ, $\bar{\Phi}$ are the background fields, and ϕ, $\bar{\phi}$ are quantum ones. The standard expression defining the quantum contribution to the effective action, $\bar{\Gamma} = \sum_{L=1}^{\infty} \hbar^L \Gamma_L$, after this splitting and introducing background fields $\Psi = m + \lambda\Phi$, $\bar{\Psi} = m + \lambda\bar{\Phi}$, takes the form:

$$e^{\frac{i}{\hbar}\tilde{\Gamma}[\Phi,\bar{\Phi}]} = \int D\phi D\bar{\phi} \exp\left(\frac{i}{2}(\phi\bar{\phi})\begin{pmatrix} \Psi & -\frac{1}{4}\bar{D}^2 \\ -\frac{1}{4}D^2 & \bar{\Psi} \end{pmatrix}\begin{pmatrix} \phi \\ \bar{\phi} \end{pmatrix} + \right.$$
$$\left. + i\sqrt{\hbar}(\frac{\lambda}{3!}\phi^3 + h.c.)\right). \tag{4.153}$$

The quadratic action of quantum superfields, used to obtain the background dependent propagator, in our case looks like

$$S^{(2)} = \frac{1}{2}(\phi\bar{\phi})\begin{pmatrix} \Psi & -\frac{1}{4}\bar{D}^2 \\ -\frac{1}{4}D^2 & \bar{\Psi} \end{pmatrix}\begin{pmatrix} \phi \\ \bar{\phi} \end{pmatrix}. \tag{4.154}$$

And the matrix Green function by definition is an operator inverse to

$$\begin{pmatrix} \Psi & -\frac{1}{4}\bar{D}^2 \\ -\frac{1}{4}D^2 & \bar{\Psi} \end{pmatrix}. \tag{4.155}$$

We can see that this Green function can be represented in the form

$$G(z_1, z_2) = \begin{pmatrix} G_{++}(z_1, z_2) & G_{+-}(z_1, z_2) \\ G_{-+}(z_1, z_2) & G_{--}(z_1, z_2) \end{pmatrix}. \tag{4.156}$$

where $+$ denotes chirality with respect to corresponding argument, and $-$ correspondingly—antichirality.

One can verify that in the Wess-Zumino model the $G(z_1, z_2)$ looks like

$$G(z_1, z_2) = \frac{1}{16}\begin{pmatrix} \bar{D}_1^2\bar{D}_2^2 G_v^\psi(z_1, z_2) & \bar{D}_1^2 D_2^2 G_v^\psi(z_1, z_2) \\ D_1^2\bar{D}_2^2 G_v^\psi(z_1, z_2) & D_1^2 D_2^2 G_v^\psi(z_1, z_2) \end{pmatrix}. \tag{4.157}$$

where $G_v^\psi(z_1, z_2) = (\Box + \frac{1}{4}\Psi\bar{D}^2 + \frac{1}{4}\bar{\Psi}D^2)^{-1}\delta^8(z_1 - z_2)$. Really, let us consider the relation

$$\frac{1}{16}\begin{pmatrix} \Psi & -\frac{1}{4}\bar{D}^2 \\ -\frac{1}{4}D^2 & \bar{\Psi} \end{pmatrix}\begin{pmatrix} \bar{D}_1^2\bar{D}_2^2 G_v^\psi(z_1, z_2) & \bar{D}_1^2 D_2^2 G_v^\psi(z_1, z_2) \\ D_1^2\bar{D}_2^2 G_v^\psi(z_1, z_2) & D_1^2 D_2^2 G_v^\psi(z_1, z_2) \end{pmatrix} = -\begin{pmatrix} \delta_+ & 0 \\ 0 & \delta_- \end{pmatrix} \tag{4.158}$$

and act on both parts of this relation with the operator

$$\begin{pmatrix} 0 & -\frac{1}{4}\bar{D}^2 \\ -\frac{1}{4}D^2 & 0 \end{pmatrix}.$$

We get the following system of equations on components of the G:

$$\Box G_{++} - \frac{1}{4}\bar{D}_1^2(\bar{\Psi}G_{-+}) = 0;$$

$$\Box G_{-+} - \frac{1}{4}D_1^2(\Psi G_{++}) = \frac{1}{16}D_1^2\bar{D}_2^2\delta^8(z_1 - z_2);$$

$$\Box G_{--} - \frac{1}{4}D_1^2(\Psi G_{+-}) = 0;$$

$$\Box G_{+-} - \frac{1}{4}\bar{D}_1^2(\bar{\Psi}G_{--}) = \frac{1}{16}\bar{D}_1^2 D_2^2\delta^8(z_1 - z_2). \tag{4.159}$$

A straightforward comparing shows that components $G_{++}, G_{+-}, G_{-+}, G_{--}$ given by (4.157) satisfy this equation. Thus, we found matrix superpropagator (4.157) which will be used for calculation of loop corrections.

Let us consider the one-loop effective action. By the definition, it is equal to

$$\Gamma^{(1)} = -\frac{i}{2}\mathrm{Tr}\log G$$

where the matrix Green function G is given by (4.157). However, a straightforward calculation of this trace is very complicated since elements of this matrix are defined in different subspaces. It follows from the definition of the generating functional (see Chap. 2) that the one-loop effective action $\Gamma^{(1)}$ can be obtained from the relation

$$e^{i\Gamma^{(1)}} = \int D\phi D\bar{\phi}\exp\left(\frac{i}{2}\begin{pmatrix}\phi\bar{\phi}\end{pmatrix}\begin{pmatrix}\Psi & -\frac{1}{4}\bar{D}^2 \\ -\frac{1}{4}D^2 & \bar{\Psi}\end{pmatrix}\begin{pmatrix}\phi \\ \bar{\phi}\end{pmatrix}\right). \tag{4.160}$$

The calculation of this path integral is essentially simplified by using the trick [27] which we discuss below and which is also applied in other theories describing dynamics of chiral superfields. We consider the theory of a real scalar superfield with the action

$$S = -\frac{1}{16}\int d^8z v D^\alpha \bar{D}^2 D_\alpha v. \tag{4.161}$$

The action is invariant under gauge transformations $\delta v = \Lambda + \bar{\Lambda}$ (here the parameter Λ is chiral, and the $\bar{\Lambda}$ is antichiral). According to Faddeev-Popov approach, the effective action W_v for this theory can be introduced as

$$e^{iW_v} = \int Dv e^{-\frac{i}{16}\int d^8z v D^\alpha \bar{D}^2 D_\alpha v}\delta(\chi). \tag{4.162}$$

Here $\delta(\chi)$ is a functional delta function, and χ is a gauge-fixing function. We choose χ in the form of column matrix

$$\chi = \begin{pmatrix}\frac{1}{4}D^2 v - \bar{\phi} \\ \frac{1}{4}\bar{D}^2 v - \phi\end{pmatrix}.$$

One should note that since D^2v and $\bar{D}^2v$ are not real superfields, we must impose two gauge-fixing conditions, and (4.162) takes the form

$$e^{iW_v} = \int Dv e^{-\frac{i}{16}\int d^8z v D^\alpha \bar{D}^2 D_\alpha v} \delta(\frac{1}{4}D^2v - \bar{\phi})\delta(\frac{1}{4}\bar{D}^2v - \phi)\det\Delta_{FP}, \quad (4.163)$$

where

$$\Delta_{FP} = \begin{pmatrix} -\frac{1}{4}\bar{D}^2 & 0 \\ 0 & -\frac{1}{4}D^2 \end{pmatrix}$$

is a Faddeev-Popov matrix. We note that W_v is constant by constructing. We multiply the left-hand and right-hand sides of (4.160) and (4.163) respectively, as a result we arrive at

$$e^{i\Gamma^{(1)}+W_v} = \int D\phi D\bar{\phi} Dv \exp\left(\frac{i}{2}(\phi\bar{\phi})\begin{pmatrix} \Psi & -\frac{1}{4}\bar{D}^2 \\ -\frac{1}{4}D^2 & \bar{\Psi} \end{pmatrix}\begin{pmatrix} \phi \\ \bar{\phi} \end{pmatrix} - \frac{i}{16}v D^\alpha \bar{D}^2 D_\alpha v\right) \times$$
$$\times \delta(\frac{1}{4}D^2v - \bar{\phi})\delta(\frac{1}{4}\bar{D}^2v - \phi)\det\Delta_{FP}. \quad (4.164)$$

Integrating over $\phi, \bar{\phi}$ with use of delta functions, we obtain

$$e^{i\Gamma^{(1)}+W_v} = \int D\phi D\bar{\phi} \exp\left(\frac{i}{2}\int d^8z v(\Box - \frac{1}{4}\Psi\bar{D}^2 - \frac{1}{4}\bar{\Psi}D^2)v\right)\det\Delta_{FP}. \quad (4.165)$$

However, W_v and $\det\Delta_{FP}$ are field-independent constants which are irrelevant for our purposes and can be omitted. We also took into account that $\frac{1}{16}\{D^2, \bar{D}^2\} - \frac{1}{8}D^\alpha\bar{D}^2 D_\alpha = \Box$, hence the one-loop effective action is equal to

$$\Gamma^{(1)} = \frac{i}{2}\text{Tr}\log(\Box - \frac{1}{4}\Psi\bar{D}^2 - \frac{1}{4}\bar{\Psi}D^2) \quad (4.166)$$

Here as usual $\Psi = m + \lambda\Phi$, $\bar{\Psi} = m + \lambda\bar{\Phi}$. This one-loop effective action can be expressed in the form of the Schwinger expansion:

$$\Gamma^{(1)} = \frac{i}{2}\text{Tr}\int_0^\infty \frac{ds}{s}\exp(is(\Box - \frac{1}{4}\Psi\bar{D}^2 - \frac{1}{4}\bar{\Psi}D^2)), \quad (4.167)$$

or, after manifest writing the trace,

$$\Gamma^{(1)} = \frac{i}{2}\int d^8z_1 d^8z_2 \int_0^\infty \frac{ds}{s}\delta^8(z_1 - z_2)\exp(is(-\frac{1}{4}\Psi\bar{D}^2 - \frac{1}{4}\bar{\Psi}D^2))e^{is\Box}\delta^8(z_1 - z_2). \quad (4.168)$$

We consider the heat kernel $\Omega(\Psi|s) = \exp(is(-\frac{1}{4}\Psi\bar{D}^2 - \frac{1}{4}\bar{\Psi}D^2)) \equiv e^{is\Delta}$. It evidently satisfies the equation

$$\frac{\partial\Omega}{\partial s} = i\Omega\Delta.$$

It turns out to be that if we calculate the Kählerian effective potential, when all super-covariant derivatives from background superfields Ψ, $\bar{\Psi}$ are omitted, this equation can be easily solved. We express Ω in the form (4.131). Then $\Omega\Delta$ is equal to

$$\begin{aligned}
\Omega\Delta = {}&-\frac{1}{4}\Psi\bar{D}^2 - \frac{1}{4}\bar{\Psi}D^2 - \\
&-\frac{1}{4}\Psi\tilde{A}\Box\bar{D}^2 - \frac{1}{4}\bar{\Psi}A\Box D^2 + \\
&+\frac{1}{4}\tilde{B}_{\dot{\alpha}}\Psi\partial^{\alpha\dot{\alpha}}D_{\alpha}\bar{D}^2 - \frac{1}{4}B_{\alpha}\bar{\Psi}\partial^{\alpha\dot{\alpha}}\bar{D}_{\dot{\alpha}}D^2 - \\
&-\frac{1}{16}\bar{\Psi}\bar{C}\bar{D}^2 D^2 - \frac{1}{16}\Psi C D^2\bar{D}^2.
\end{aligned} \tag{4.169}$$

Comparing coefficients at analogous derivatives in $\frac{1}{i}\frac{\partial\Omega}{\partial s}$ and $\Omega\Delta$ we get the following system of equations

$$\begin{aligned}
\dot{A} &= -\Psi C; \\
\dot{B}^{\alpha} &= 2i\,\tilde{B}_{\dot{\alpha}}\Psi\partial^{\alpha\dot{\alpha}}; \\
\dot{C} &= -\bar{\Psi} - \bar{\Psi}A\Box.
\end{aligned} \tag{4.170}$$

The system for $\tilde{A}$, $\tilde{B}$, $\tilde{C}$ has the analogous form with changing $\Psi \to \bar{\Psi}$, $A \to \tilde{A}$, etc. Here a dot denotes $\frac{1}{i}\frac{\partial}{\partial s} \equiv \frac{\partial}{\partial\tilde{s}}$, and $\tilde{s} = is$. Since $\Omega|_{s=0} = 1$, and all terms in expansion of Ω (4.131) are evidently linearly independent, natural initial conditions are

$$A = \tilde{A} = B^{\alpha} = \tilde{B}_{\dot{\alpha}} = C = \tilde{C}|_{s=0} = 0. \tag{4.171}$$

We find that the system of equations for B^{α} and $\tilde{B}_{\dot{\alpha}}$ is closed (it is isolated from the whole system (4.170)) and homogeneous. The initial conditions (4.171) imply that its only solution is zero, B^{α}, $\tilde{B}_{\dot{\alpha}} = 0$. The remaining system for A and C (and the analogous one for $\tilde{A}$ and $\tilde{C}$) is actually a standard system of usual first-order differential equations. Its solution looks like

$$C = -\sqrt{\frac{\bar{\Psi}\Box}{\Psi}}(A_0^1\exp(i\omega s) - A_0^2\exp(-i\omega s))$$

$$A = A_0^1\exp(i\omega s) + A_0^2\exp(-i\omega s) - \frac{1}{\Box}. \tag{4.172}$$

Here $\omega = \sqrt{\Psi\bar{\Psi}\Box}$. Imposing initial conditions (4.171) allows to fix coefficients A_0^1, A_0^2. As a result we get

$$C = -\sqrt{\frac{\bar{\Psi}}{\Psi\Box}} \sinh(is\sqrt{\Psi\bar{\Psi}\Box})$$

$$A = \frac{1}{\Box}[\cosh(is\sqrt{\Psi\bar{\Psi}\Box}) - 1]. \tag{4.173}$$

Since A is symmetric with respect to change $\Psi \to \bar{\Psi}$ we find that $A = \tilde{A}$. We note that only A and $\tilde{A}$ contribute to the trace in (4.168) since they are accompanied by $D^2\bar{D}^2$ and $\bar{D}^2D^2$, unique combinations of supercovariant derivatives which can yield a non-zero trace in the superspace. Therefore the one-loop Kählerian contribution to effective action is equal to

$$\Gamma_K^{(1)} = i \int d^4x d^4\theta \int_0^\infty \frac{d\tilde{s}}{\tilde{s}} \frac{1}{\Box}[\cosh(\tilde{s}\sqrt{\Psi\bar{\Psi}\Box}) - 1]U^{(0)}(x, x'; s)|_{x=x'}. \tag{4.174}$$

Here $U^{(0)}(x, x'; s)$ is given by (4.125). This function satisfies the equation (see Sect. 4.6):

$$\Box^n U^{(0)}(x, x'; s)|_{x=x'} = -i(\frac{\partial}{\partial\tilde{s}})^n \frac{1}{16\pi^2\tilde{s}^2}.$$

We expand (4.174) into power series:

$$\frac{1}{\Box}[\cosh(\tilde{s}\sqrt{\Psi\bar{\Psi}\Box}) - 1] = \sum_{n=0}^\infty \tilde{s}^{2n+2} \frac{(\Psi\bar{\Psi})^{n+1}}{(2n+2)!}\Box^n.$$

And

$$\Gamma_K^{(1)} = i \int d^4x d^4\theta \int_0^\infty \frac{d\tilde{s}}{\tilde{s}} \sum_{n=0}^\infty \tilde{s}^{2n+2} \frac{(\Psi\bar{\Psi})^{n+1}}{(2n+2)!}\Box^n U^{(0)}(x, x'; s)|_{x=x'} =$$

$$= -i \int d^4x d^4\theta \int_0^\infty \frac{d\tilde{s}}{\tilde{s}} \sum_{n=0}^\infty \tilde{s}^{2n+2} \frac{(\Psi\bar{\Psi})^{n+1}}{(2n+2)!}(\frac{\partial}{\partial\tilde{s}})^n \frac{-i}{16\pi^2\tilde{s}^2} =$$

$$= -\frac{1}{32\pi^2} \int d^8z \int_{L^2}^\infty \frac{d\tilde{s}}{\tilde{s}^2} \sum_{n=0}^\infty (-1)^n \frac{(\tilde{s}\Psi\bar{\Psi})^{n+1}(n+1)!}{(2n+2)!}. \tag{4.175}$$

Since the integral diverges at the lower limit, we introduce the cutoff L^2 for the regularization. Then, we make the change $\tilde{s}\Psi\bar{\Psi} = t$ so that t is dimensionless. As a result, the one-loop Kählerian effective potential takes the form

$$K^{(1)} = -\frac{1}{32\pi^2}\Psi\bar{\Psi} \int_{\Psi\bar{\Psi}L^2}^\infty \frac{dt}{t^2} \sum_{n=0}^\infty \frac{(n+1)!t^{n+1}(-1)^n}{(2n+2)!}. \tag{4.176}$$

Then, $\sum_{n=0}^\infty \frac{(n+1)!t^{n+1}(-1)^n}{(2n+2)!} = t \int_0^1 du e^{-\frac{t}{4}(1-u^2)}$. Hence

$$K^{(1)} = -\frac{1}{32\pi^2} \Psi \bar\Psi \int_{\Psi\bar\Psi L^2}^{\infty} \frac{dt}{t} \int_0^1 du e^{-\frac{t}{4}(1-u^2)}. \tag{4.177}$$

At $L^2 \to 0$ this integral tends to

$$K^{(1)} = -\frac{1}{32\pi^2} \Psi \bar\Psi \log(\mu^2 L^2) - \frac{1}{32\pi^2} \Psi \bar\Psi (\log \frac{\Psi \bar\Psi}{\mu^2} - \xi), \tag{4.178}$$

where ξ is some constant which can be absorbed into redefinition of μ. We can add the counterterm $\frac{1}{32\pi^2} \Psi \bar\Psi \log(\mu^2 L^2)$ to cancel the divergence. Such a counterterm corresponds to a usual wave function renormalization by the rule

$$\Phi \to Z^{1/2}\Phi; \quad Z = 1 + \frac{\lambda^2}{32\pi^2} \log(\mu^2 L^2). \tag{4.179}$$

And the renormalized Kählerian effective potential is

$$K_{ren}^{(1)} = -\frac{1}{32\pi^2} \Psi \bar\Psi (\log \frac{\Psi \bar\Psi}{\mu^2} - \xi). \tag{4.180}$$

Another way for calculating of the Kählerian effective potential consists in summarizing of contributions from supergraphs given by Fig. 4.5. The sum of these contributions looks like [75]

$$K^{(1)} = \int \frac{d^4k}{(2\pi)^4} \int d^4\theta_1 \ldots d^4\theta_{2n} \sum_{n=1}^{\infty} \frac{1}{2n} \left(\frac{\Psi \bar\Psi}{k^4}\right)^n \frac{D^2}{4}\delta_{12} \frac{\bar{D}^2}{4}\delta_{23} \ldots \frac{D^2}{4}\delta_{2n-1,2n} \frac{\bar{D}^2}{4}\delta_{2n,1} \tag{4.181}$$

which after D-algebra transformations and summation looks like

$$K^{(1)} = \mu^\epsilon \int \frac{d^{4-\epsilon}k}{(2\pi)^{4-\epsilon}} \frac{1}{2k^2} \log(1 + \frac{\Psi \bar\Psi}{k^2}), \tag{4.182}$$

where we carried out a dimensional regularization by introducing the parameter ϵ. Integrating, one obtains

$$K^{(1)} = \frac{1}{32\pi^2} [\frac{\Psi \bar\Psi}{\epsilon} - \Psi \bar\Psi \log \frac{\Psi \bar\Psi}{e\mu^2}] \tag{4.183}$$

where $e = \exp(1)$. A subtraction of the divergence and a redefinition of μ leads to the result (4.180).

As for the one-loop auxiliary fields effective potential $F^{(1)}$, its calculation is much more involved. In [27], the lower (four-derivative) contribution to it has been found. Only in [76], after a rather tricky procedure the complete result for it has been obtained in terms of dilogarithms of some combinations of spinor supercovariant derivatives of background superfields.

Now we turn to calculating of the chiral effective potential. It differs from zero for massless theories. Really, as it was noted by West [67], the mechanism of arising chiral corrections is the following one. If the theory describes dynamics of chiral and antichiral superfields, then quantum correction of the form

$$\int d^8 z f(\Phi)(-\frac{D^2}{4\Box}) g(\Phi) \tag{4.184}$$

can be rewritten as

$$\int d^6 z f(\Phi) g(\Phi). \tag{4.185}$$

Here we used properties $\int d^8 z = \int d^6 z(-\frac{\bar{D}^2}{4})$ and $\bar{D}^2 D^2 \Phi = 16\Box\Phi$ (the last identity is valid for any chiral superfield Φ), and $f(\Phi)$, $g(\Phi)$ are arbitrary functions of the chiral superfield Φ. However, the presence of the factor $\Box^{-1}$ is characteristic for massless theories, whereas in massive theories we have $(\Box - m^2)^{-1}$ instead of $\Box^{-1}$, and this mechanism of arising contributions to the chiral effective potential does not work. We note that this situation is rather generic, that is, the perturbative contributions to the chiral effective potential can arise only if all propagators are massless (except of very rare situations where the integral over massive propagators completely factorizes out giving only a constant with no dependence on external momenta). Actually, namely this effect is crucial to prove the Goldstone theorem in noncommutative superfield theories [68] which is related with the statement of absence of Φ^2 corrections in theories with chiral self-couplings.

In the case of the massless theory we can find the matrix Green function (4.157) exactly: first,

$$G_v^\psi(z_1, z_2) = (\Box + \frac{1}{4}\Psi\bar{D}^2)^{-1}\delta^8(z_1 - z_2) = \frac{1}{\Box_1}\delta^8(z_1 - z_2) -$$

$$- \frac{1}{\Box_1}\Psi(z_1)\frac{\bar{D}_1^2}{4\Box_1}\delta^8(z_1 - z_2). \tag{4.186}$$

The higher terms in this expansion are equal to zero because they are proportional to $\bar{D}^2\Psi = 0$ or $\bar{D}^4 = 0$. It follows straightforwardly from this expression that $\mathrm{Tr}\ln G_v^\psi = 0$, therefore, the one-loop effective potential in the Wess-Zumino model identically vanishes. The components of matrix superpropagator (4.157) look like

$$G_{++} = 0; \quad G_{+-} = G_{-+}^* = \frac{\bar{D}_1^2 D_2^2}{16\Box}\delta^8(z_1 - z_2);$$

$$G_{--} = -\frac{D_1^2}{4\Box_1}[\Psi(z_1)\frac{\bar{D}_1^2 D_2^2}{16\Box}\delta^8(z_1 - z_2)]. \tag{4.187}$$

Fig. 4.7 The contribution to the two-loop chiral effective potential in the Wess-Zumino model

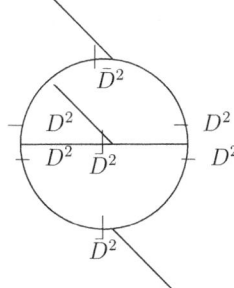

Here the $*$ symbol is for a complex conjugated term. We note that the background chiral superfield Ψ is not constant, otherwise the expression proportional to $D^2\Psi$ will yield a singularity $\frac{0}{0}$ [72]. The only two-loop contribution to the chiral effective potential is given by the supergraph depicted at Fig. 4.7.

External lines stand for Ψ. It is convenient to use the chiral representation where $\Phi(z) = \Phi(x, \theta)$, and $\bar{D}_{\dot\alpha} = -\frac{\partial}{\partial\theta^{\dot\alpha}}$. In the massless case, $\Psi = \lambda\Phi$.

The contribution of the supergraph given in the Fig. 4.7 looks like

$$
I = \frac{\lambda^5}{12} \int \frac{d^4 p_1 d^4 p_2}{(2\pi)^8} \frac{d^4 k d^4 l}{(2\pi)^8} \int d^4\theta_1 d^4\theta_2 d^4\theta_3 d^4\theta_4 d^4\theta_5 \Phi(-p_1, \theta_4)\Phi(-p_2, \theta_5) \times
$$

$$
\times \; \Phi(p_1 + p_2, \theta_3) \frac{1}{k^2 l^2 (k + p_1)^2 (l + p_2)^2 (l + k)^2 (l + k + p_1 + p_2)^2} \times
$$

$$
\times \; \delta_{13} \frac{\bar{D}_3^2}{4} \delta_{32} \frac{D_1^2 \bar{D}_4^2}{16} \delta_{14} \frac{D_4^2}{4} \delta_{42} \frac{D_1^2 \bar{D}_5^2}{16} \delta_{15} \frac{D_5^2}{4} \delta_{52}. \tag{4.188}
$$

After D-algebra transformations this expression takes the form

$$
I = \frac{\lambda^5}{12} \int \frac{d^4 p_1 d^4 p_2}{(2\pi)^8} \frac{d^4 k d^4 l}{(2\pi)^8} \int d^2\theta \Phi(-p_1, \theta)\Phi(-p_2, \theta)\Phi(p_1 + p_2, \theta) \times
$$

$$
\times \; \frac{k^2 p_2^2 + l^2 p_1^2 + 2(kl)(p_1 p_2)}{k^2 l^2 (k + p_1)^2 (l + p_2)^2 (l + k)^2 (l + k + p_1 + p_2)^2}. \tag{4.189}
$$

Here we used the identity $\int d^4\theta = \int d^2\theta(-\frac{1}{4}\bar{D}^2)$ and took into account that $\bar{D}^2 D^2 \Phi(p, \theta) = -16 p^2 \Phi(p, \theta)$.

By the definition, the effective potential is the effective Lagrangian for superfields slowly varying in space-time. Let us study the behavior of the expression (4.189) in this limit. The contribution (4.189) can be expressed as

$$
I = \frac{\lambda^5}{12} \int d^2\theta \int \frac{d^4 p_1 d^4 p_2}{(2\pi)^8} \Phi(-p_1, \theta)\Phi(-p_2, \theta)\Phi(p_1 + p_2, \theta) S(p_1, p_2). \tag{4.190}
$$

Here p_1, p_2 are external momenta, and

$$S(p_1, p_2) = \int \frac{d^4k d^4 l}{(2\pi)^8} \frac{k^2 p_2^2 + l^2 p_1^2 + 2(kl)(p_1 p_2)}{k^2 l^2 (k + p_1)^2 (l + p_2)^2 (l + k)^2 (l + k + p_1 + p_2)^2}.$$
(4.191)

After the Fourier transform Eq. (4.190) becomes

$$I = \frac{\lambda^5}{12} \int d^2\theta \int d^4x_1 d^4x_2 d^4x_3 \int \frac{d^4 p_1 d^4 p_2}{(2\pi)^8} \Phi(x_1, \theta)\Phi(x_2, \theta) \times$$
$$\times \ \Phi(x_3, \theta) \exp[i(-p_1 x_1 - p_2 x_2 + (p_1 + p_2)x_3)]S(p_1, p_2). \qquad (4.192)$$

Since superfields in the case under consideration are assumed to be slowly varying in the space-time we can put $\Phi(x_1, \theta)\Phi(x_2, \theta)\Phi(x_3, \theta) \simeq \Phi^3(x_1, \theta)$. As a result one gets

$$I = \frac{\lambda^5}{12} \int d^2\theta \int d^4x_1 d^4x_2 d^4x_3 \int \frac{d^4 p_1 d^4 p_2}{(2\pi)^8} \Phi^3(x_1, \theta) \times$$
$$\times \ \exp[i(-p_1 x_1 - p_2 x_2 + (p_1 + p_2)x_3)]S(p_1, p_2). \qquad (4.193)$$

Integrating over $d^4x_2 d^4x_3$, we obtain delta-functions $\delta(p_2)\delta(p_1 + p_2)$. Hence the Eq. (4.192) takes the form

$$I = \frac{\lambda^5}{12} \int d^2\theta \int d^4x_1 \Phi^3(x_1, \theta) S(p_1, p_2)|_{p_1, p_2=0}. \qquad (4.194)$$

Therefore the final result for the two-loop contribution to the chiral (holomorphic) effective potential looks like

$$W^{(2)} = \frac{\lambda^5}{2(16\pi^2)^2} \zeta(3)\Phi^3(z), \qquad (4.195)$$

where we took into account that (cf. [25, 77])

$$S(p_1, p_2)|_{p_1, p_2=0}$$
$$= \int \frac{d^4k d^4 l}{(2\pi)^8} \frac{k^2 p_2^2 + l^2 p_1^2 + 2(kl)(p_1 p_2)}{k^2 l^2 (k + p_1)^2 (l + p_2)^2 (l + k)^2 (l + k + p_1 + p_2)^2}|_{p_1=p_2=0} =$$
$$= \frac{6}{(4\pi)^4} \zeta(3). \qquad (4.196)$$

We see that the correction (4.195) is finite and does not require any renormalization. Actually, it follows from [77] that a whole class of two-loop finite supergraphs yields contributions proportional to $\zeta(3)$. In principle, all finite two-loop contributions to

Fig. 4.8 One-loop contribution to the chiral effective potential in the $\mathcal{N} = 1$ SYM with chiral matter

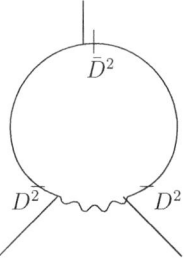

the chiral effective potential in the SYM theory which we consider below have just this structure hence they are proportional to $\zeta(3)$.

The situation in $\mathcal{N} = 1$ super-Yang-Mills theory with (massless) chiral matter, however, possesses some peculiarities. First of all, in this theory the one-loop contribution to the chiral effective potential is nontrivial. It is described by the supergraph depicted at Fig. 4.8.

It was shown in [26] that the contribution of this diagram, after simple D-algebra transformations, looks like

$$W^{(1)} = -\frac{1}{16\pi^2} C_0 \lambda_{dec} g^2 (T^I)^d_a (T^I)^e_b \int d^6z \, \Phi^a(z) \Phi^b(z) \Phi^c(z), \quad (4.197)$$

where

$$C_0 = \int_0^1 d\alpha \frac{\ln \alpha(1 - \alpha)}{1 - \alpha(1 - \alpha)} \quad (4.198)$$

is a finite constant. There is no other one-loop contributions to the chiral effective potential.

As for the two-loop order, both finite and divergent two-loop chiral contributions in this theory are possible. The finite ones are depicted at Fig. 4.9, and the divergent ones—at Fig. 4.10. We use the Feynman gauge for propagators of the gauge superfield, to avoid the infrared singularities. Contributions of all finite diagrams to the chiral effective potential can be shown to be proportional to $\frac{\zeta(3)}{(4\pi)^4} \Phi^3$ (the corresponding integral over momenta looks similarly to (4.196) and yields just this result) being analogous to the case of Wess-Zumino model and matching the class of contributions described in [77].

As an example of the divergent contribution we will consider the supergraph given by Fig. 4.10a. After simple D-algebra transformation, it is proportional to

$$\int d^2\theta \int \frac{d^4 p_1 d^4 p_2}{(2\pi)^8} (p_1 + p_2)^2 \int \frac{d^4 k d^4 l}{(2\pi)^8} \frac{1}{k^2(k + p_1)^2(k + p_2)^2} \times$$

$$\times \frac{1}{l^2(l + p_1 + p_2)^2} \Phi(-p_1, \theta) \Phi(-p_2, \theta) \Phi(p_1 + p_2, \theta). \quad (4.199)$$

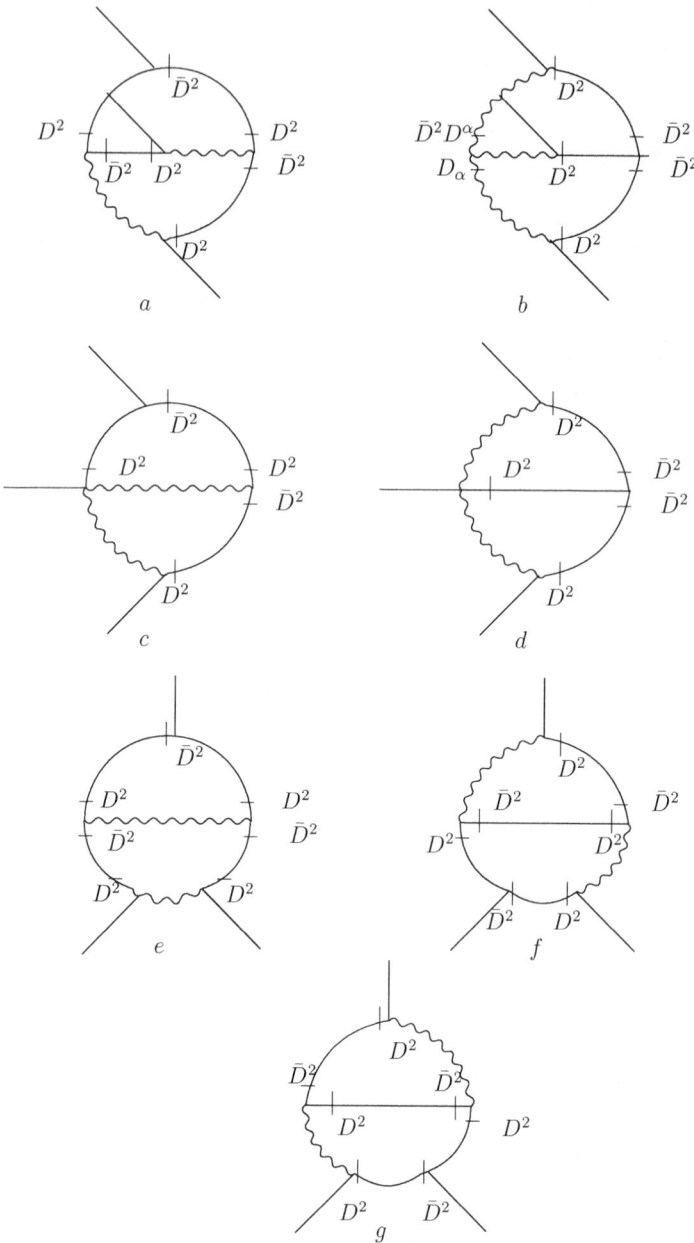

Fig. 4.9 Finite contributions to the two-loop chiral effective potential in $\mathcal{N} = 1$ SYM theory

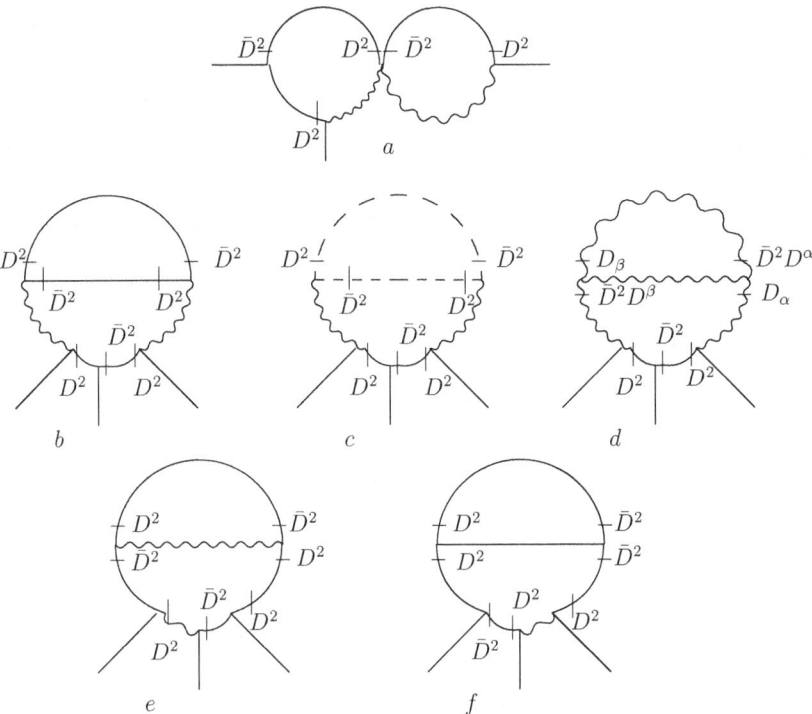

Fig. 4.10 Divergent contributions to the two-loop chiral effective potential in $\mathcal{N} = 1$ SYM theory

To obtain the low-energy leading contribution we must consider the limit at p_1, $p_2 \to 0$. It is known [26] that

$$\lim_{p_1, p_2 \to 0} (p_1 + p_2)^2 \int \frac{d^4 k}{(2\pi)^4} \frac{1}{k^2 (k + p_1)^2 (k + p_2)^2}$$
$$= \frac{1}{16\pi^2} \int_0^1 d\alpha \frac{\log[\alpha(1 - \alpha)]}{1 - \alpha(1 - \alpha)} = \frac{C_0}{16\pi^2},$$

where C_0 is the finite constant given by (4.198). The integral over l diverges, and after the dimensional regularization it is equal to

$$\int \frac{d^{4-\epsilon} l}{(2\pi)^{4-\epsilon}} \frac{1}{l^2 (l + p_1 + p_2)^2} = \frac{1}{16\pi^2} (\frac{2}{\epsilon} + \log \frac{(p_1 + p_2)^2}{\mu^2}).$$

After cancellation of the divergence with the help of the appropriate one-loop counterterm and returning to the coordinate space, we see that expression (4.199) for superfields slowly varying in space-time takes the form

$$\int d^6 z \frac{1}{(16\pi^2)^2} \Phi^2(z) \log(-\frac{\Box}{\mu^2})\Phi(z). \tag{4.200}$$

Carrying out the calculations for all supergraphs above we conclude that the two-loop low-energy leading chiral correction in the $\mathcal{N} = 1$ SYM theory with chiral matter is nonlocal. Its complete form, after subtracting of the corresponding counterterms, is

$$
\begin{aligned}
\tilde{\Gamma}_c^{(2)} &= \frac{g^2}{(4\pi)^4} \{\zeta(3)[6^5 g^2 (\lambda_{imn}(T^K T^I T^K)_k^n (T^I)_j^m + \lambda_{lnp}(T^I T^K)_k^p (T^I)_i^l (T^K)_j^n + \\
&\quad + \lambda_{imn}(T^I T^K)_k^n \{T^K, T^I\}_j^m) + 4 \times 6^3 (T^I T^J)_i^l (T^J)_k^m (\lambda_{lmp}(T^I)_j^p + \\
&\quad + \lambda_{ljp}(T^I)_m^p)] + 3 \times 6^4 [\frac{\lambda_{iml} g^2}{3!} (\frac{1}{64} - 4)(T^K)_b^a (T^I)_d^c (T^K)_j^m (T^I)_k^l \times \\
&\quad \times [(T^L),(T^N)]_a^b [(T^L),(T^N)]_c^d + \frac{\lambda_{rnl}\lambda_{rns}\lambda_{skm}}{(3!)^3} (T^I)_i^l (T^I)_j^m + \\
&\quad + \lambda_{snp} g^2 (T^K)_i^s (T^K)_k^p (T^I)_m^n (T^I)_j^m + \\
&\quad + \lambda_{irp} g^2 (T^I)_n^m (T^K)_m^n (T^I)_k^p (T^K)_j^r](1-\gamma)C_0 + \\
&\quad + 2 \times 6^4 \frac{\lambda_{iml}}{2!3!} g^2 (T^K)_p^s (T^K)_k^p (T^I)_s^l (T^I)_j^m (1-\frac{\gamma}{2}C_0)\} \times \\
&\quad \times \int d^6 z \Phi^i(z)\Phi^j(z)\Phi^k(z) - \\
&\quad - \frac{2g^4}{(4\pi)^4} 6^4 \frac{\lambda_{iml}}{2!3!} (T^K)_p^s (T^K)_k^p (T^I)_s^l (T^I)_j^m \int d^6 z \Phi^i(z)\Phi^j(z)[\ln\left(-\frac{\Box}{\mu^2}\right)]\Phi^k(z) - \\
&\quad - 3 \times 6^4 \{\int d^2\theta \frac{d^4 p_1 d^4 p_2}{(2\pi)^8} [\frac{\lambda_{iml} g^2}{3!} (\frac{1}{64} - 4)(T^K)_b^a (T^I)_d^c (T^K)_j^m (T^I)_k^l \times \\
&\quad \times [(T^L),(T^N)]_a^b [(T^L),(T^N)]_c^d + \lambda_{irp} g^4 (T^I)_n^m (T^K)_m^n (T^I)_k^p (T^K)_j^r + \\
&\quad + \frac{\lambda_{rnl}\lambda_{rns}\lambda_{skm} g^2}{(3!)^3} (T^I)_i^l (T^I)_j^m + \lambda_{smp} g^4 (T^I)_n^m (T^I)_j^n (T^K)_k^p (T^K)_i^s]\} \times \\
&\quad \times \int_0^1 d\beta d\alpha [\ln \frac{p_1^2\beta(1-\beta) + p_2^2\alpha(1-\alpha) - 2p_1 p_2\alpha\beta}{\mu^2(1-\alpha-\beta)}] \times \tag{4.201} \\
&\quad \times \frac{(p_1+p_2)^2}{p_1^2\beta(1-\beta) + p_2^2\alpha(1-\alpha) - 2p_1 p_2\alpha\beta} \Phi^i(-p_1,\theta)\Phi^j(-p_2,\theta)\Phi^k(p_1+p_2,\theta).
\end{aligned}
$$

This result is found for the SYM theory with an arbitrary non-Abelian gauge group. In the particular case of the $SU(N)$ group, the result reduces to

$$\tilde{\Gamma}_c^{(2)} = a_1 \lambda_{ijk} g^4 \int d^6z \Phi^i(z) \Phi^j(z) \Phi^k(z) +$$

$$+ a_2 \lambda_{ijk} g^4 \int d^6z \Phi^i(z) \Phi^j(z) [\ln(-\frac{\Box}{\mu^2})] \Phi^k(z) +$$

$$+ \lambda_{rnj} \lambda_{rns} \lambda_{ski} g^2 \Big(a_3 \int d^6z \Phi^i(z) \Phi^j(z) \Phi^k(z) +$$

$$+ a_4 \int d^6z \Phi^i(z) \Phi^j(z) [\ln(-\frac{\Box}{\mu^2})] \Phi^k(z) \Big). \tag{4.202}$$

Here the constants a_1, a_2, a_3, a_4 are equal to

$$a_1 = \frac{1}{(4\pi)^4} \{ \zeta(3)[6^5(-\frac{N-1}{2N^2} + \frac{1}{4}(1-\frac{1}{N})^2 + \frac{1}{4}(N+1-\frac{2}{N})) +$$

$$= 2 \times 6^3 (1-\frac{1}{N})^2] + 3 \times 6^4 [\frac{1}{3!} \frac{N-1}{2}(\frac{1}{256} - 1) +$$

$$+ \frac{(N^2-1)(N-1)}{4N^2} + \frac{N-1}{4N}](1-\frac{\gamma}{2})C_0 +$$

$$+ 6^3 \frac{(N^2-1)(N-1)}{4N^2}(1-\frac{\gamma}{2})C_0\} -$$

$$- \frac{C}{(4\pi)^4} 3 \times 6^4 \{ \frac{1}{3!}(\frac{1}{256} - 1)\frac{N-1}{2} + \frac{(N-1)(N^2-1)}{4N^2} + \frac{N-1}{4N} \};$$

$$a_2 = -6^3 \frac{1}{(4\pi)^4} \frac{(N-1)(N^2-1)}{4N^2} - \tag{4.203}$$

$$- \frac{C_0}{(4\pi)^4} 3 \times 6^4 \{ \frac{1}{3!}(\frac{1}{256} - 1)\frac{N-1}{2} + \frac{(N-1)(N^2-1)}{4N^2} + \frac{N-1}{4N} \};$$

$$a_3 = (1-\frac{1}{N})\frac{9}{(4\pi)^4}(C + (1-\frac{\gamma}{2})C_0);$$

$$a_4 = (1-\frac{1}{N})\frac{9}{(4\pi)^4} C_0,$$

and

$$C = \int_0^1 d\beta d\alpha [\ln \frac{p_1^2 \beta(1-\beta) + p_2^2 \alpha(1-\alpha) - 2p_1 p_2 \alpha\beta}{(p_1+p_2)^2(1-\alpha-\beta)} - 1] \times$$

$$\times \frac{(p_1+p_2)^2}{p_1^2 \beta(1-\beta) + p_2^2 \alpha(1-\alpha) - 2p_1 p_2 \alpha\beta}|_{p_1,p_2 \to 0}. \tag{4.204}$$

We note that nonlocal corrections in the SYM theory arise also in the pure gauge sector [78].

This nonlocality is a natural consequence of the presence of UV divergences in a theory with massless fields. However, treating the problem of the possible renormalization of the chiral effective potential (to the best of our knowledge, it was not discussed yet in scientific literature), one must note that the divergence in the two-loop chiral effective potential is caused by the one-loop divergent contributions to the two- and three-point functions involving the gauge superfield. In the maximal, $\mathcal{N} = 4$ supersymmetric case, these divergences must mutually cancel (for the two-point functions, both of gauge and chiral superfields, such a cancellation has been explicitly shown in [23], see also [2]). It allows to conclude that in the "complete", $\mathcal{N} = 4$ SYM theory, no such a nonlocality arises. Another way for interpretation of this situation can be related with the concept of the Wilsonian effective action [79] whose use naturally introduces an infrared cutoff parameter Λ with a subsequent replacement of the nonlocal factor $\ln(-\frac{\Box}{\mu^2})$ in (4.202) by a constant $\ln(\frac{\Lambda^2}{\mu^2})$, thus, this expression acquires the usual local form of the chiral effective potential.

Chiral contributions to the effective action arise also in other theories describing dynamics of chiral superfields. For example, in a general chiral superfield theory (see the next section), the leading chiral contribution is also a chiral effective potential, while in the higher-derivative field theory the leading chiral contribution is of second order in space-time derivatives of a chiral superfield (see Sect. 4.10), and these corrections are finite.

We conclude that the presence of quantum contributions to the chiral effective Lagrangian is indeed quite characteristic for theories including chiral superfields, as it was claimed in the seminal paper [67].

4.9 General Chiral Superfield Model

In the previous section, we have studied the Wess-Zumino model, that is, the simplest example of a theory describing a dynamics of the chiral superfield. The natural question is the possibility for constructing of the most generic model for this superfield. Such a model naturally arises within the context of the superstring theory, from which viewpoint, low-energy effective field theory models represent themselves as theories where the integration over massive string modes is carried out, and the extra dimensions of the ten-dimensional background manifold, where the dynamics of the superstring takes place, are compactified, so, the structure of this manifold has the form $M^4 \times K$ where M^4 is the usual four-dimensional Minkowski space, and K is a some six-dimensional compact space. As a result, after reducing to M^4, one arrives at the theory of chiral superfields described by the action [1]:

$$S[\Phi, \bar{\Phi}] = \int d^8 z K(\Phi^i, \bar{\Phi}^i) + (\int d^6 z W(\Phi^i) + h.c.). \qquad (4.205)$$

We can use matrix notations via introduction of column vector $\vec{\Phi} = \{\Phi^i\}$, after which the consideration in the case of several chiral superfields is analogous to the case of one chiral superfield (some extensions of this model involving the gauge superfields are discussed in [80]). We can consider this theory for arbitrary functions K and W.

By this definition, there is no higher derivatives in the classical action. Therefore the theory with the action (4.205) is the most general theory without higher derivatives describing a dynamics of a chiral superfield. There are a lot of phenomenological applications of this model in string theory (see [80] and references therein). In a general case this theory is non-renormalizable. However, non-renormalizable theories are naturally treated within the framework of the effective field theory approach [81]. Therefore all integrals over momenta effectively involve natural cutoff through the condition $p \ll M_{String}$ where p is momentum, and $M_{String} = 10^{17} GeV \sim 10^{-2} M_{Pl}$ is a characteristic string mass.

The effective action in the theory, just as in the Wess-Zumino model, can be presented as a series in supercovariant derivatives $D_A = (\partial_a, D_\alpha, \bar{D}_{\dot\alpha})$ in the form (4.147)–(4.151) being described by the same objects, that is, Kählerian effective potential K_{eff}, auxiliary fields' effective potential $\mathbf{F}_{eff}$ which we do not consider in this section, and the chiral effective potential W_{eff}.

The one-loop contribution to the effective action is totally determined by the quadratic part of expansion of $\frac{1}{\hbar} S[\bar\Phi + \sqrt{\hbar}\bar\phi, \Phi + \sqrt{\hbar}\phi]$ in quantum fields $\phi, \bar\phi$ which looks like

$$S_2 = \frac{1}{2} \int d^8 z \left(\phi \; \bar\phi \right) \begin{pmatrix} K_{\Phi\Phi} & K_{\Phi\bar\Phi} \\ K_{\Phi\bar\Phi} & K_{\bar\Phi\bar\Phi} \end{pmatrix} \begin{pmatrix} \phi \\ \bar\phi \end{pmatrix} + [\int d^6 z \frac{1}{2} W'' \phi^2 + h.c.] \quad (4.206)$$

and defines the propagators, and the higher terms of this expansion define the vertices. Here $K_{\Phi\bar\Phi} = \frac{\partial^2 K(\bar\Phi, \Phi)}{\partial\Phi\partial\bar\Phi}$, $K_{\Phi\Phi} = \frac{\partial^2 K(\bar\Phi, \Phi)}{\partial\Phi^2}$, etc, $W'' = \frac{d^2 W}{d\Phi^2}$.

The corresponding matrix Green function is again represented by the form (4.156). In the case when all derivatives of superfields Φ, $\bar\Phi$ are omitted (that is just the case of the Kählerian effective potential), it satisfies the equation

$$\begin{pmatrix} W'' & -\frac{1}{4} K_{\Phi\bar\Phi} \bar{D}^2 \\ -\frac{1}{4} K_{\Phi\bar\Phi} D^2 & \bar{W}'' \end{pmatrix} \begin{pmatrix} G_{++}(z_1, z_2) & G_{+-}(z_1, z_2) \\ G_{-+}(z_1, z_2) & G_{--}(z_1, z_2) \end{pmatrix} = - \begin{pmatrix} \delta_+ & 0 \\ 0 & \delta_- \end{pmatrix}.$$
$$(4.207)$$

The solution of this equation looks like

$$G = \frac{1}{K_{\Phi\bar\Phi}^2 \Box - W'' \bar{W}''} \begin{pmatrix} \bar{W}'' & \frac{1}{4} K_{\Phi\bar\Phi} \bar{D}^2 \\ \frac{1}{4} K_{\Phi\bar\Phi} D^2 & W'' \end{pmatrix} \begin{pmatrix} \delta_+ & 0 \\ 0 & \delta_- \end{pmatrix}. \quad (4.208)$$

Now we turn to studying of quantum contributions to the Kählerian effective potential depending only on superfields Φ, $\bar\Phi$ but not on their derivatives.

The one-loop Feynman supergraphs contributing to the Kählerian effective potential are given by Fig. 4.11.

Here, bold external lines correspond to alternating W'' and $\bar{W}''$. The bold internal lines are the background dependent $\langle \phi\bar\phi \rangle$-propagators of the form

$$G_0 \equiv \langle \phi\bar\phi \rangle = -i \frac{\bar{D}^2 D^2}{16 K_{\Phi\bar\Phi} \Box} \delta^8(z_1 - z_2). \quad (4.209)$$

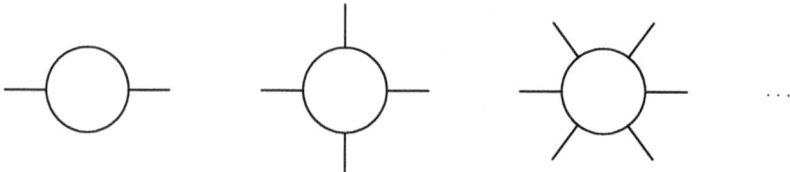

Fig. 4.11 One-loop supegraphs contributing to Kählerian effective potential

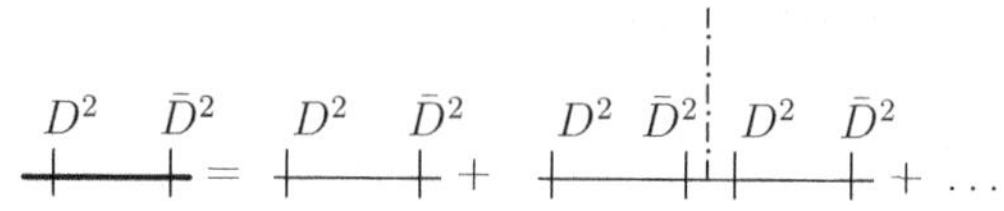

Fig. 4.12 Obtaining the background dependent propagator (4.209) through summation

$$D^2 \quad \bar{D}^2 \quad D^2 \quad \bar{D}^2 \quad D^2 \quad \bar{D}^2 \quad D^2 \quad \bar{D}^2$$

Fig. 4.13 A standard link in a diagram contributing to the one-loop effective potential

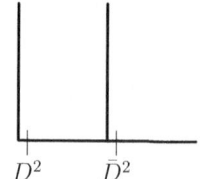

$$D^2 \qquad \bar{D}^2$$

We note that this propagator is valid for calculating not only of the Kählerian effective potential but also of the chiral one. It can be obtained in two ways. The first way consists in a summation over insertions (see Fig. 4.12).

Here dashed-and-dotted vertical line is the external field $K_{\Phi\bar{\Phi}} - 1$, and the horizontal simple line is the usual propagator $-\bar{D}_1^2 D_2^2 \frac{\delta^8(z_1 - z_2)}{16\Box}$: either if the external field is constant or if it is chiral, we find that the sum of the contributions (with integrals in internal points are assumed where it is necessary) above is

$$\sum_{n=0}^{\infty} -\frac{\bar{D}^2 D^2}{16\Box}[-(K_{\Phi\bar{\Phi}} - 1)\frac{\bar{D}^2 D^2}{16\Box}]^n \delta^8(z_1 - z_2) = -\frac{\bar{D}_1^2 D_2^2}{16 K_{\Phi\bar{\Phi}}(z_1)\Box} \delta^8(z_1 - z_2), \tag{4.210}$$

which proves (4.209).

In another manner, the result (4.209) can be proved through the definition of the new chiral field $\phi' = K_{\Phi\bar{\Phi}}\phi$ (indeed, in this case $K_{\Phi\bar{\Phi}}$ is either constant or chiral, so, ϕ' can be assumed to be chiral as well, so that the free action acquires the form $S_f = \int d^8 z \phi' \bar{\phi}$ which yields $\langle \phi'(z_1)\bar{\phi}(z_2)\rangle = -\frac{\bar{D}_1^2 D_2^2}{16\Box}\delta^8(z_1 - z_2)$, which, in its part, implies (4.209).

A supergraph of the structure depicted at Fig. 4.11, with $2n$ legs represents itself as a ring containing n links of the form given by Fig. 4.13.

Carrying out the calculations in the same way as in the previous section (see [82, 85] for the details), we find that the total contribution of all these diagrams after D-algebra transformations, summation, integration over momenta and subtraction of divergences is equal to

$$K^{(1)} = -\frac{1}{32\pi^2} \, \text{tr} \, \frac{W'' \bar{W}''}{K_{\Phi\bar{\Phi}}^2} \ln \left(\frac{W'' \bar{W}''}{\mu^2 K_{\Phi\bar{\Phi}}^2} \right), \qquad (4.211)$$

where tr denotes trace of product of the given matrices. This form is more convenient for the analysis of a many-field generalization of our model than that one given in [62, 82], it is evident that this result reduces to the known expression for the Wess-Zumino model (4.180), for the choice $W'' = \lambda\Phi$.

Let us consider the chiral (holomorphic) effective potential $W_{eff}(\Phi)$. The mechanism of its arising is just the same than in Wess-Zumino model. We note again that the chiral contributions to effective action can be generated by supergraphs containing massless propagators only. To find such corrections to effective action we put $\bar{\Phi} = 0$ in Eq. (4.206). Therefore here and further all derivatives of K, W and $\bar{W}$ in (4.206) will be taken at $\bar{\Phi} = 0$. We refer to the theory as to the massless one if $W''|_{\Phi=0} = 0$. Further we consider only a massless theory and follow [83].

To construct supergraphs which yield chiral contributions one splits the action (4.206) into sum of the free part and vertices of interaction. As a free part we take the action $S_0 = \int d^8 z \, K_{\Phi\bar{\Phi}} \phi\bar{\phi}$. And the term $S[\bar{\phi}, \phi, \Phi] - S_0$ will be treated as vertices where $S[\bar{\phi}, \phi, \Phi]$ is given by Eq. (4.206). Our purpose is to find the first leading contribution to $W_{eff}(\Phi)$. The straightforward inspection shows that there is no one-loop contributions to the chiral effective potential, just as in the Wess-Zumino model. As we will show, in a whole analogy with it, chiral contributions in our theory begin at two loops. Therefore we keep in Eq. (4.206) only the terms of second, third and fourth orders in quantum fields which are sufficient in the two-loop approximation. Also, the vertex $K_{\Phi\Phi\bar{\phi}}\phi^2$ evidently yields zero contribution to the chiral effective potential. As for vertices $K_{\bar{\Phi}\bar{\Phi}}\bar{\phi}^2$, they cannot contribute to the chiral effective potential as it follows from a straightforward calculation of numbers of quantum fields ϕ, $\bar{\phi}$ (which should be equal) and D-factors [83].

As a result we find that the only two-loop supergraph contributing to chiral effective potential is that one given at Fig. 4.14.

In Fig. 4.14, bold external lines are W'', and bold internal lines are propagators $\langle \phi\bar{\phi} \rangle$ (4.209), where, however, the superfield $K_{\Phi\bar{\Phi}}$ is not restricted to be a constant more but depends on a background chiral superfield. We note that this supergraph is

Fig. 4.14 A contribution to the two-loop chiral effective potential

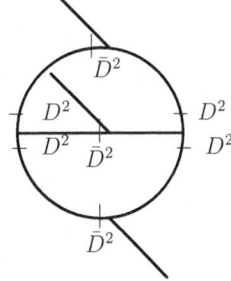

analogous to that one contributing to the chiral effective potential in the Wess-Zumino model.

After D-algebra transformations and loop integrations completely analogous to those ones carried out in the previous section we find that two-loop contribution to the chiral effective potential in this model looks like

$$W^{(2)} = \frac{1}{2(16\pi^2)^2}\zeta(3)\lambda^2\left\{\frac{W''(z)}{K^2_{\Phi\bar{\Phi}}(z)}\right\}^3. \tag{4.212}$$

One reminds that $\lambda = \bar{W}'''(\bar{\Phi})|_{\bar{\Phi}=0}$ and $K_{\Phi\bar{\Phi}}(z) = \frac{\partial^2 K(\bar{\Phi},\Phi)}{\partial\Phi\partial\bar{\Phi}}|_{\bar{\Phi}=0}$ here. We see that the correction (4.212) is finite and does not require renormalization in any case despite the theory is non-renormalizable in general case. In the case of the Wess-Zumino model it reduces to (4.195).

We note that the calculation of two-loop Kählerian effective potential can be carried out with help of matrix superpropagator (4.208). The results are given in [62, 82].

A coupling of the general chiral superfield model to the gauge superfield can be introduced as well, and corresponding calculations are performed along the same lines as here. The details of calculations are given in [70]. Qualitatively the one- and two-loop Kählerian effective potentials in this theory are rather similar to the results obtained for the Wess-Zumino model and supersymmetric QED which have been discussed in this chapter.

Now let us consider some phenomenological applications of the theory characterized by the action (4.205). Let us suppose that the column vector $\vec{\Phi}$ describes two superfields: the light (massless) one ϕ and the heavy one Φ, so, $\vec{\Phi} = \begin{pmatrix} \phi \\ \Phi \end{pmatrix}$. For this case, we find the one-loop effective action and eliminate heavy superfields with use of their effective equations of motion. As a result we arrive at the effective action of light superfields. There is a decoupling theorem [71, 84] according to which this effective action after redefining of parameters, such as fields, masses, couplings, can be expressed in the form of a sum of effective action of the theory obtained from initial one by putting heavy fields to zero and terms proportional to different powers of $\frac{1}{M}$ where M is mass of heavy superfield (which in the case under consideration is put, by phenomenological reasons, to be equal to the characteristic string mass M_{String} [80, 82]).

We study such a theory in the one-loop approximation. The low-energy leading one-loop contribution to effective action is given by (4.211). In principle, one can proceed with diagonalization of the matrices $K_{\Phi\bar{\Phi}} = \frac{\partial^2 K}{\partial\Phi^i\partial\bar{\Phi}^j}$ and $W'' = \frac{\partial^2 W}{\partial\Phi^i\partial\Phi^j}$. However, within this review let us restrict ourselves by a qualitative description of the situation only.

We consider, as a didactic example, the minimal theory whose action is slightly different from that one used [85], so,

$$K = \phi\bar{\phi} + \Phi\bar{\Phi}; \quad W = \frac{M}{2}\Phi^2 + \frac{\lambda}{2}\phi\Phi^2 + \frac{g}{3!}\phi^3. \quad (4.213)$$

Let us, for the sake of the simplicity, consider the situation when the heavy superfield Φ is a purely quantum one, and the light superfield ϕ is a purely background one. In this case the quadratic action of Φ looks like

$$S_2 = \int d^8z\,\Phi\bar{\Phi} + (\frac{1}{2}\int d^6z(M + \lambda\phi)\Phi^2 + h.c.) \quad (4.214)$$

It is evident that this action identically replays the quadratic action of the quantum fields in the usual Wess-Zumino model after the background-quantum splitting (besides of (4.206), see also the previous section of this review and [27]). So, one can repeat all calculations performed for the Wess-Zumino model, and the renormalized one-loop contribution to the low-energy effective action is given by the result (4.211), where $K(\Phi, \bar{\Phi}) = \Phi\bar{\Phi}$. Explicitly, the one-loop Kählerian effective potential in our theory is

$$K_\Phi^{(1)} = -\frac{1}{32\pi^2}(M + \lambda\phi)(M + \lambda\bar{\phi})\ln\frac{(M + \lambda\phi)(M + \lambda\bar{\phi})}{\mu^2}. \quad (4.215)$$

We expand it in power series in $\frac{1}{M}$, with introducing a new parameter $\mu' = \alpha\mu$, where α is a some real number:

$$K_\Phi^{(1)} = -\frac{1}{32\pi^2}\lambda^2\phi\bar{\phi}\ln\frac{M^2}{\mu'^2} + O(\frac{1}{M}). \quad (4.216)$$

This expression involves the term $\ln\frac{M^2}{\mu^2}$. It is clear that this term is not suppressed at $M \to \infty$ but instead of this, increases as M grows. Moreover, if one suggests that the light field ϕ is not purely external but also possesses a nontrivial quantum dynamics, the contribution to one-loop Kählerian effective potential generated by Feynman diagrams constructed of light propagators is

$$K_\phi^{(1)} = -\frac{1}{32\pi^2}g^2\phi\bar{\phi}\ln\frac{g^2\phi\bar{\phi}}{\mu^2}, \quad (4.217)$$

and if we fix the normalization parameter as $\mu = \alpha M$, we see that in this case, instead of the contribution (4.216), the (4.217) will grow as $M \to \infty$. The same situation can be verified as well for other models involving light and heavy fields in different situations (presence of nonminimal couplings, nontrivial quantum and background parts both for light and heavy fields etc.) [85].

To argue the existence of the contributions increasing with growth of M in a general case, one can notice as well the following facts. First, if the theory involves a cubic self-coupling of the light (massless) chiral superfield ϕ with a coupling g, as a consequence of integrating over light quantum fields the one-loop effective action will involve the term (4.217). The situation does not essentially change if this self-coupling is described by the generic function $V(\phi)$, instead of the cubic one, and the corresponding contribution to the one-loop effective action can be obtained through a straightforward replacement $\phi \to V''(\phi)$ in (4.217). At the same time, the sector in which the light superfield ϕ is coupled to the heavy one Φ, in the usual case when the theory display divergences, certainly, among other terms, will yield the contribution $b \int d^8z \phi \bar{\phi} \ln \frac{M^2}{\mu^2}$, where b is also finite (this contribution is always present if the corresponding supergraphs involve massive propagators), as we already showed. Therefore, either this contribution increases with growth of M, or, if one fixes the parameter μ imposing the condition $M = \mu$ (or $M = b\mu$, with a constant b), the one-loop contribution of the light fields only takes the form $c \int d^8z g^2 \phi \bar{\phi} \ln \frac{g^2 \phi \bar{\phi}}{M^2}$ which increases with growing of M. Actually we demonstrated that such increasing contributions emerge in each theory involving heavy fields, if the result displays a dependence on μ (i.e. if the theory involves divergences), which is common if the theory does not involve higher derivatives or extended supersymmetry.

4.10 The Higher-Derivative Chiral Superfield Models

The next generalization for the models of the chiral superfield consists in including higher derivatives. Originally, a higher-derivative extension of a supersymmetric theory has been introduced for the purposes of the regularization in [86]. Besides of this, the higher-derivative terms have been implemented in gravity, first in [87] where it was shown to achieve a renormalizable gravity. Moreover, it was found in [88] that the higher-derivative additive modifications of the gravity action arise due to the presence of the conformal anomaly of matter fields in a curved space. In [89], a supersymmetric analogue of this anomaly and the corresponding additive modification of the supergravity action were obtained, and in [63] the quantum dynamics of the dilaton superfield was studied.

Actually, the higher-derivative field theories are studied in different contexts, including different gravity modifications which are intensively applied to explain the cosmic acceleration [91], and the Horava-Lifshitz model of gravity [92]. In the context of supersymmetry, the new wave of an interest to higher-derivative superfield theories was recently called by the paper [93]. Treating the notorious problem of arising of the ghosts whose presence is typical for the higher-derivative field theories, we note that an attempt to solve this problem was carried out in [94], where the hypothesis that contributions of ghosts can be decoupled in certain cases (so-called "benign ghosts") was discussed. At the same time, it is natural to treat the higher-derivative theories as effective models for studying of the low-energy effects, see the discussion in [95].

We start with the simplest example of a higher-derivative superfield theory [96]:

$$S[\Phi, \bar{\Phi}] = \int d^8z \Phi \Box \bar{\Phi} + (\int d^6z W(\Phi) + h.c.). \qquad (4.218)$$

Here Φ is a chiral superfield, and $W(\Phi)$ is an arbitrary function. A particular form of this action was studied in [85], for the case $W(\Phi) = \Lambda e^{3\Phi}$, it corresponds to the four-dimensional anomaly-modified dilaton supergravity in the infrared limit, where, in particular, all derivative-dependent coupling terms of the classical action simply vanish. This action, being reduced to the components, contains terms of the fourth order in space-time derivatives. We will refer to this theory as to the minimal higher-derivative theory.

The effective action $\Gamma[\Phi, \bar{\Phi}]$ is defined as a generating functional of the one-particle-irreducible vertex Green functions:

$$e^{i\Gamma[\Phi,\bar{\Phi}]} = \int D\phi D\bar{\phi} \exp(iS[\Phi + \phi, \bar{\Phi} + \bar{\phi}])|_{1PI}. \qquad (4.219)$$

Here the Φ, $\bar{\Phi}$ are background (classical) fields and ϕ, $\bar{\phi}$ are quantum fields. As usual, we can represent the structure of the effective action in this theory in the form given by (4.147–4.151).

To obtain the one-loop effective action, one should expand the r.h.s. of the equation (4.219) up to the second order in the quantum superfields ϕ, $\bar{\phi}$ (cf. [35]). As a result, the one-loop effective action turns out to be defined from the expression:

$$e^{i\Gamma^{(1)}[\Phi,\bar{\Phi}]} = \int D\phi D\bar{\phi} \exp(i[\int d^8 z \phi \Box \bar{\phi} + (\frac{1}{2}\int d^6 z W''(\Phi)\phi^2 + h.c.)]), \qquad (4.220)$$

and the $\Gamma^{(1)}[\Phi, \bar{\Phi}]$ can be represented in the form of the functional supertrace:

$$\Gamma^{(1)}[\Phi, \bar{\Phi}] = \frac{i}{2}\text{Tr}\ln\begin{pmatrix} W'' & -\Box\frac{\bar{D}^2}{4} \\ -\Box\frac{D^2}{4} & \bar{W}'' \end{pmatrix}. \qquad (4.221)$$

Just as it occurs in the Wess-Zumino model, the elements of this matrix are defined in different subspaces of the superspace and mix chiralities. To find a more simple equivalent form of $\Gamma^{(1)}[\Phi, \bar{\Phi}]$, we again, as in the previous sections, use the trick based on the Faddeev-Popov methodology.

Let us consider the free higher-derivative theory of the real scalar superfield whose action is

$$S_v = -\frac{1}{16}\int d^8 z v D^\alpha \bar{D}^2 D_\alpha \Box v. \qquad (4.222)$$

This action is evidently invariant under the usual gauge transformations $\delta v = \Lambda + \bar{\Lambda}$, where Λ is a chiral superfield, and $\bar{\Lambda}$ is an antichiral one. Following general prescriptions of the Faddeev-Popov method, one can define the effective action W_v of this theory as

$$e^{iW_v} = \int Dv \exp(-\frac{i}{16} \int d^8z v D^\alpha \bar{D}^2 D_\alpha \Box v) \delta(\frac{1}{4} D^2 v - \bar{\phi}) \delta(\frac{1}{4} \bar{D}^2 v - \phi) \Delta_{FP},$$

$$(4.223)$$

where the $\frac{1}{4} D^2 v - \bar{\phi}$, $\frac{1}{4} \bar{D}^2 v - \phi$ play the role of the gauge fixing functions, the ϕ, $\bar{\phi}$ are the same as in (4.220), and the Δ_{FP} is the Faddeev-Popov determinant. One should notice that the W_v is a constant.

Then, let us multiply correspondingly the left-hand and right-hand sides of the expressions (4.220) and (4.223). The functional integration over ϕ, $\bar{\phi}$ is straightforward, and after omitting irrelevant constants, the one-loop effective action takes the form

$$\Gamma^{(1)} = \frac{i}{2} \text{Tr} \ln(\Box^2 - \frac{1}{4} W''(\Phi) \bar{D}^2 - \frac{1}{4} \bar{W}''(\bar{\Phi}) D^2). \qquad (4.224)$$

So, the expression for the one-loop effective action is simplified crucially.

Thus, we face the problem of calculating of trace of the logarithm of the higher-derivative operator. The most convenient way to do it is based on the use of the proper-time representation (see Sects. 4.6 and 4.8):

$$\Gamma^{(1)} = \frac{i}{2} \text{Tr} \int \frac{ds}{s} \exp[is(\Box^2 + \frac{1}{4} \Psi \bar{D}^2 + \frac{1}{4} \bar{\Psi} D^2)]. \qquad (4.225)$$

Here we denoted $W''(\Phi) = -\Psi$, $\bar{W}''(\bar{\Phi}) = -\bar{\Psi}$ for the convenience. One should remind that Ψ is a chiral superfield, and $\bar{\Psi}$ is an antichiral one.

Disregarding the terms involving the space-time derivatives of Φ, $\bar{\Phi}$, which will not contribute to lower orders of the derivative expansion of the effective action, we can rewrite this expression as

$$\Gamma^{(1)} = \frac{i}{2} \int d^8z_1 \int \frac{ds}{s} \exp[is(\frac{1}{4} \Psi \bar{D}^2 + \frac{1}{4} \bar{\Psi} D^2)] e^{is\Box^2} \delta^8(z_1 - z_2)|_{z_1=z_2}. \quad (4.226)$$

Now, let us proceed in a way similar to that one used in Sect. 4.8. As a first step, we introduce operators

$$\Delta = \frac{1}{4} \Psi \bar{D}^2 + \frac{1}{4} \bar{\Psi} D^2; \quad \Omega(\Psi, \bar{\Psi}, s) = e^{is\Delta}, \qquad (4.227)$$

where Ω can be again expanded in the form (4.131) and satisfies the superfield heat conductivity equation

$$\frac{1}{i} \frac{d\Omega}{ds} = \Omega\Delta. \qquad (4.228)$$

The initial condition is evidently $\Omega|_{s=0} = 1$, hence $A(s = 0) = \tilde{A}(s = 0) = B_\alpha(s = 0) = \tilde{B}_{\dot\alpha}(s = 0) = C(s = 0) = \tilde{C}(s = 0) = 0$. The system involving these coefficients turns out to be exactly the same as in the Wess-Zumino case (see Sect. 4.8), hence the coefficients A and $\tilde{A}$ (they are again the only ones contributing to the one-loop effective potential) reproduce the results obtained in the Wess-Zumino model [27]. In our case, unlike Sect. 4.8, we give these coefficients up to the fourth order in the spinor supercovariant derivatives of superfields:

$$
\begin{aligned}
A(s) + \tilde{A}(s) = {} & \frac{2}{\square}[\cosh(\tilde{s}U) - 1] + \\
& + \tilde{s}\frac{D^2\Psi\bar{D}^2\bar{\Psi}}{64\square}(\tilde{s}\cosh(\tilde{s}U) - \frac{1}{U}\sinh(\tilde{s}U)) + \\
& + \frac{\tilde{s}}{64U^2}[\bar{\Psi}\bar{D}^2\bar{\Psi}(D^\alpha\Psi)(D_\alpha\Psi) + \Psi D^2\Psi(\bar{D}_{\dot{\alpha}}\bar{\Psi})(\bar{D}^{\dot{\alpha}}\bar{\Psi})] \times \\
& \times (\frac{1}{3}\tilde{s}^2 U\sinh(\tilde{s}U) - \tilde{s}\cosh(\tilde{s}U) + \frac{1}{U}\sinh(\tilde{s}U)) + \\
& + \frac{\tilde{s}}{256}(D^\alpha\Psi)(D_\alpha\Psi)(\bar{D}_{\dot{\alpha}}\bar{\Psi})(\bar{D}^{\dot{\alpha}}\bar{\Psi})[\frac{1}{2}\tilde{s}^3\cosh(\tilde{s}U) - \frac{5}{3}\frac{\tilde{s}^2}{U}\sinh(\tilde{s}U) + \\
& + \frac{7}{2U^2}(\tilde{s}\cosh(\tilde{s}U) - \frac{1}{U}\sinh(\tilde{s}U))].
\end{aligned}
\tag{4.229}
$$

Here $\tilde{s} = is$, $U = \sqrt{\Psi\bar{\Psi}\square}$. The higher orders in supercovariant derivatives of Ψ, $\bar{\Psi}$ in principle also can be found, however, the complete result would be extremely cumbersome.

The one-loop effective action can be expressed as

$$
\Gamma^{(1)} = -\frac{i}{2}\int d^4\theta d^4x_1 \int \frac{ds}{s}[A(s) + \tilde{A}(s)]e^{is\square^2}\delta^4(x_1 - x_2)|_{x_1=x_2}. \tag{4.230}
$$

This result differs from that one obtained in the Wess-Zumino model since the d'Alembertian operators from the expansion of $A(s) + \tilde{A}(s)$, will act not on the usual function $e^{is\square}\delta^8(z_1 - z_2)$, as it occurs in that case, but on the function $e^{is\square^2}\delta^8(z_1 - z_2)$.

It remains to substitute (4.229) into (4.230) and, afterwards, to expand it in the power series in $\square$. The Kählerian contribution to the one-loop effective action is given by the first line of (4.229), i.e.

$$
\Gamma^{(1)}_K = -i\int d^4\theta d^4x_1 \int \frac{ds}{s}\frac{1}{\square}[\cosh(\tilde{s}U) - 1]e^{is\square^2}\delta^4(x_1 - x_2)|_{x_1=x_2}, \tag{4.231}
$$

which, after expanding in series in $\square$ yields

$$
\Gamma^{(1)}_K = \int d^4\theta d^4x_1 \int \frac{dt}{t}\sum_{n=0}^{\infty}\frac{1}{(2n+2)!}(t^2\Psi\bar{\Psi})^{n+1}\square^n e^{-t\square^2}\delta^4(x_1 - x_2)|_{x_1=x_2},
$$
$$
\tag{4.232}
$$

where we carried out the Wick rotation $s = it$ (with $t = -\tilde{s}$) and $x_0 = ix_{0E}$ for convenience.

It is convenient to split the indices n into odd, $n = 2l + 1$ and even, $n = 2l$, ones. As a result, we can write $K^{(1)}$ as

$$\Gamma_K^{(1)} = \int d^4\theta d^4x_1 \int \frac{dt}{t} \sum_{l=0}^{\infty} \left[\frac{1}{(4l+2)!} (t^2 \Psi \bar{\Psi})^{2l+1} \Box^{2l} + \frac{1}{(4l+4)!} (t^2 \Psi \bar{\Psi})^{2l+2} \Box^{2l+1} \right] \times$$
$$\times e^{-t\Box^2} \delta^4(x_1 - x_2)|_{x_1=x_2}. \tag{4.233}$$

Now, let us consider the structure $\Box^n e^{-t\Box^2} \delta^4(x_1 - x_2)|_{x_1=x_2}$. It is clear that the function $V(t; x_1, x_2) = e^{-t\Box^2} \delta^4(x_1 - x_2)$ which we will call the free heat kernel, satisfies the equation

$$\Box^2 V(t; x_1, x_2) = -\frac{d}{dt} V(t; x_1, x_2), \tag{4.234}$$

hence

$$\Box^{2l} V(t; x_1, x_2) = (-\frac{d}{dt})^l V(t; x_1, x_2); \quad \Box^{2l+1} V(t; x_1, x_2) = (-\frac{d}{dt})^l \Box V(t; x_1, x_2). \tag{4.235}$$

In this subsection, the above expressions will be considered only in the limit $x_1 = x_2$. One can find that

$$V(t; x_1, x_2)|_{x_1=x_2} = \int \frac{d^4k}{(2\pi)^4} e^{-tk^4} = \frac{1}{32\pi^2 t};$$
$$\Box V(t; x_1, x_2)|_{x_1=x_2} = \int \frac{d^4k}{(2\pi)^4} (-k^2) e^{-tk^4} = -\frac{1}{32\pi^{3/2} t^{3/2}}, \tag{4.236}$$

therefore

$$\Box^{2l} V(t; x_1, x_2)|_{x_1=x_2} = (-\frac{d}{dt})^l \frac{1}{32\pi^2 t} = \frac{(-1)^l l!}{32\pi^2 t^{l+1}};$$
$$\Box^{2l+1} V(t; x_1, x_2)|_{x_1=x_2} = (-\frac{d}{dt})^l (-\frac{1}{32\pi^{3/2} t^{3/2}}) = -\frac{(-1)^{l+1}(2l+1)!!}{32\pi^{3/2} 2^l t^{l+3/2}}. \tag{4.237}$$

Replacing all this into (4.233), we arrive at

$$\Gamma_K^{(1)} = \frac{1}{32\pi^2} \int d^8z \int dt \sum_{l=0}^{\infty} (-1)^l \left[t^{3l} \frac{l!(\Psi\bar{\Psi})^{2l+1}}{(4l+2)!} - t^{3l+3/2} \frac{(\Psi\bar{\Psi})^{2l+2}}{(4l+4)!} \frac{\sqrt{\pi}(2l+1)!!}{2^l} \right]. \tag{4.238}$$

To simplify this expression, let us make the change $t(\Psi\bar{\Psi})^{2/3} = u$ (we note that u is dimensionless). We find

$$\Gamma_K^{(1)} = \frac{1}{32\pi^2} \int d^8z (\Psi\bar{\Psi})^{1/3} \int du \sum_{l=0}^{\infty} \left[\frac{(-1)^l u^{3l} l!}{(4l+2)!} - \right.$$
$$\left. - \frac{(-1)^l u^{3l+3/2}}{(4l+4)!} \frac{\sqrt{\pi}(2l+1)!!}{2^l} \right], \tag{4.239}$$

so the one-loop Kählerian effective potential is

$$K^{(1)} = \frac{c_0}{32\pi^2} \int d^8z (\Psi\bar{\Psi})^{1/3}, \tag{4.240}$$

where

$$c_0 = \int du \sum_{l=0}^{\infty} \left[\frac{(-1)^l u^{3l} l!}{(4l+2)!} - \frac{(-1)^l u^{3l+3/2}}{(4l+4)!} \frac{\sqrt{\pi}(2l+1)!!}{2^l} \right] \tag{4.241}$$

is a finite constant. It is easy to see that the result for dilaton supergravity [63], being a particular case of this result, is explicitly reproduced.

Now, let us calculate the one-loop auxiliary fields' effective action. To do it, let us consider all derivative dependent terms in (4.229). After their expansion in power series in $\Box$, we find

$$\Gamma_F^{(1)} = -i \int d^4\theta d^4x_1 \int \frac{dt}{t} \sum_{n=0}^{\infty} \left[\frac{D^2\Psi \bar{D}^2\bar{\Psi}}{64} t^{2n+4} (\Psi\bar{\Psi})^{n+1} \left[\frac{1}{(2n+2)!} - \frac{1}{(2n+3)!} \right] + \right.$$
$$+ \frac{1}{64} [\bar{\Psi}\bar{D}^2\bar{\Psi}D^\alpha\Psi D_\alpha\Psi + h.c.] t^{2n+4} (\Psi\bar{\Psi})^n \left[\frac{1}{3(2n+1)!} - \frac{1}{(2n+2)!} + \frac{1}{(2n+3)!} \right] +$$
$$+ \frac{1}{256} D^\alpha\Psi D_\alpha\Psi \bar{D}_{\dot{\alpha}}\bar{\Psi} \bar{D}^{\dot{\alpha}}\bar{\Psi} t^{2n+6} (\Psi\bar{\Psi})^n \times$$
$$\times \left[\frac{1}{2(2n)!} - \frac{5}{3(2n+1)!} + \frac{7}{2(2n+2)!} - \frac{7}{2(2n+3)!} \right] \times$$
$$\left. \times \Box^n f e^{-t\Box^2} \delta^4(x_1 - x_2)|_{x_1=x_2}. \right. \tag{4.242}$$

Then, we apply the same scheme as above. By its essence, this expression looks like

$$\Gamma_F^{(1)} = i \int d^4\theta d^4x_1 \int \frac{dt}{t} \sum_{n=0}^{\infty} A_n(\Psi, \bar{\Psi}, t)\Box^n e^{-t\Box^2} \delta^4(x_1 - x_2)|_{x_1=x_2}. \tag{4.243}$$

Here A_n are some functions of fields whose explicit form can be read off from (4.242). After carrying out the transformations we used above, we find the auxiliary fields' effective potential to be

$$F^{(1)} = C_1 \frac{\bar{D}^2 \bar{\Psi} D^2 \Psi}{\Psi \bar{\Psi}} + C_2 [\bar{\Psi} \bar{D}^2 \bar{\Psi} D^\alpha \Psi D_\alpha \Psi + h.c.] \frac{1}{(\Psi \bar{\Psi})^2} +$$

$$+ C_3 D^\alpha \Psi D_\alpha \Psi \bar{D}_{\dot{\alpha}} \bar{\Psi} \bar{D}^{\dot{\alpha}} \bar{\Psi} \frac{1}{(\Psi \bar{\Psi})^2}, \tag{4.244}$$

where C_1, C_2, C_3 are some numbers. We note that, in principle, this form can be predicted without explicit calculations. Indeed, this form should involve exactly two D_α derivatives and two $\bar{D}_{\dot{\alpha}}$ derivatives. Also, by dimensional reasons, the numbers of fields Ψ (and similarly $\bar{\Psi}$) should be equal in a numerator and a denominator of any contribution to this expression, which also must be symmetric with respect to the change $\Psi \to \bar{\Psi}$. Hence we can have only the terms listed in the expression above.

To close the consideration of the one-loop effective action for this model, let us discuss the one-loop chiral contributions to the effective action. It is clear that they differ from zero only if $\bar{W}''(\bar{\Phi})|_{\bar{\Phi}=0} = \Lambda \neq 0$. It means that the Λ is related with the mass of the theory. Thus, the theory is massive, hence there is no contributions to the chiral effective potential. It was showed in [90] that there are two possible types of chiral contributions in this theory, that is, $f_1(\Psi) \Box \Psi$ and $f_2(\Psi) \partial^m \Psi \partial_m \Psi$, however, the first of them can be reduced to the second one through integration by parts. Hence, we have the only possible form for the formally chiral quantum contribution to the effective Lagrangian, that is,

$$\mathcal{L}_c^{(1)} = f(\Psi) \partial^m \Psi \partial_m \Psi. \tag{4.245}$$

By dimensional reasons, one can write this expression as

$$\mathcal{L}_c^{(1)} = a f(\Psi/\Lambda) \Psi^{-5/3} \partial^m \Psi \partial_m \Psi, \tag{4.246}$$

where a is a some number. Actually, it is convenient to rewrite this expression in terms of Φ and suggest that $W''(\Phi)|_{\Phi=0} = \Lambda \neq 0$ as well, in this case we can suggest e.g. $W''(\Phi) = \Lambda e^{3\Phi}$, as it occurs in the dilaton supergravity, we have

$$\mathcal{L}_c^{(1)} = f(e^{3\Phi}) e^\Phi \partial^m \Phi \partial_m \Phi. \tag{4.247}$$

The explicit form of the function $f(e^{3\Phi})$ is given in [90]. Actually this function is a linear combination of some exponentials $e^{n\Phi}$ with different values of n.

We note however that this expression, first, does not contribute to the chiral effective potential since it involves derivatives of background superfields, second, can be treated (and perhaps it is more natural) as a contribution to the auxiliary fields' effective potential although it is chiral, since, being expressed in the form of the integral over the whole superspace, it contains no nonlocalities since it looks like $\int d^8 z f(\Psi) D^\alpha \Psi D_\alpha \Psi$, with $f(\Psi)$ is a some function of Ψ only, with no derivatives or nonlocal factors like $\Box^{-1}$.

Now, it is very instructive to discuss an application of the proper-time approach to a slightly different form of the higher-derivative theory whose kinetic term is

$$S_K = \int d^8 z \Phi (\square - M^2) \bar{\Phi}. \tag{4.248}$$

This kinetic term is equivalent to the one of the Wess-Zumino model with a higher-derivative regulator [86]. The importance of theories with such a kinetic term follows from the observation made in [93] where the higher-derivative superfield theory namely with this kinetic term has been shown to be classically equivalent to the chiral superfield theory which does not involve higher derivatives, but, instead of this, describes a dynamics of a set of chiral superfields. The key idea of [93] consists in introducing new chiral fields χ and $\tilde{\chi}$, which are related with Φ and $\bar{D}^2 \Phi$ via some linear transformation, which is, however, singular at $M = 0$. So, the studies carried out in the paper [93] are applicable only for the theory with $M \neq 0$.

Let us calculate the one-loop low-energy effective action for a theory with this kinetic term. The application of the proper-time method in this case turns out to be more complicated than for the theory (4.218). While the calculation of the Schwinger coefficients $A(s)$ and $\tilde{A}(s)$ is the same as above, the analogue of the free heat kernel function $V(t; x_1, x_2)$ can be shown to be equal to

$$V(s; x_1, x_2) = e^{-s(\square^2 - M^2 \square)} \delta^4 (x_1 - x_2). \tag{4.249}$$

However, even the evaluation of the case $x_1 = x_2$, which is only interesting for us in the one-loop approximation, is a nontrivial problem which admits a simple solution only for a very large mass M. Let us proceed with this calculation.

After the Fourier transform and the Wick rotation, the function $V(t; x_1, x_2)|_{x_1 = x_2}$ looks like

$$I(s) \equiv V(s; x_1, x_2)|_{x_1 = x_2} = \int \frac{d^4 k}{(2\pi)^4} e^{-s(k^4 + k^2 M^2)}. \tag{4.250}$$

Changing variables, $k^2 = u$, we find

$$I(s) = \frac{1}{16\pi^2} e^{\frac{sM^4}{4}} \int_0^\infty du u e^{-s(u + \frac{M^2}{2})^2}. \tag{4.251}$$

Replacing then $u + \frac{M^2}{2} = u'$ and integrating over u where it is possible, we find

$$I(s) = \frac{1}{32\pi^2 s} - \frac{M^2}{32\pi^2} e^{\frac{sM^4}{4}} \int_{M^2/2}^\infty du e^{-su^2}. \tag{4.252}$$

We find that this expression for the heat kernel function can be expressed through the probability integral function

$$\Phi(x) = \frac{2}{\sqrt{\pi}} \int_0^x dt e^{-t^2}. \tag{4.253}$$

The presence of such a function seems to make impossible finding the explicit one-loop Kählerian potential in the general case. It is clear that $\Phi(x \to \infty) \to 1$. Indeed,

$$I(s) = \frac{1}{16\pi^2} \left(\frac{1}{2s} - \frac{M^2}{2} e^{sM^2/4} \left(\int_0^\infty du e^{-su^2} - \int_0^{M^2/2} du e^{-su^2} \right) \right) \quad (4.254)$$

Substituting $su^2 = w^2$, we get

$$I(s) = \frac{1}{16\pi^2} \left(\frac{1}{2s} - \frac{M^2}{2} e^{sM^2/4} \left(\frac{1}{2} \sqrt{\frac{\pi}{s}} - \frac{1}{\sqrt{s}} \int_0^{M^2\sqrt{s}/2} dw e^{-w^2} \right) \right) =$$

$$= \frac{1}{16\pi^2} \left[\frac{1}{2s} - \frac{M^2}{2} e^{sM^2/4} \frac{1}{2} \sqrt{\frac{\pi}{s}} (1 - \Phi(M^2\sqrt{s}/2)) \right]. \quad (4.255)$$

To evaluate this expression, we employ the asymptotics of the probability integral $\Phi(y)$ at large arguments [97]:

$$\Phi(y)|_{y\to\infty} = 1 - \frac{1}{\pi} e^{-y^2} \sum_{k=0}^\infty \frac{(-1)^k \Gamma(k + \frac{1}{2})}{y^{2k+1}}. \quad (4.256)$$

We find that the term with $k = 0$ identically cancels the "usual" term $\frac{1}{2s}$ in (4.254). Taking into account only the $M \to \infty$ dominant term (remind that the limit of very high masses was studied earlier in [85]), one finds

$$I(s) = \frac{1}{16\pi^2 s^2 M^4}, \quad (4.257)$$

which differs from the case $M = 0$ considered earlier where the analogue of this function was proportional to $\frac{1}{s}$. One could note that such behavior of the heat kernel seems to be similar to that one occurring in the Wess-Zumino model [27, 72]. Nevertheless, the presence of a large mass in the denominator gives a hope that the corrections to the effective action will be suppressed in a $M \to \infty$ limit.

To give a description of the principal difference of new theory, we restrict ourselves only to calculating the Kählerian effective potential in a new theory. Let its action be

$$S[\Phi, \bar{\Phi}] = \int d^8 z \Phi(\Box - M^2)\bar{\Phi} + \left(\int d^6 z W(\Phi) + h.c. \right). \quad (4.258)$$

Here M is a large parameter related to the physical mass. To simplify the calculation of the effective action in the theory, it is natural to represent the one-loop contribution in the form of a functional integral over the unique unconstrained real scalar field as we have done in the Sect. 4.8 for the Wess-Zumino model. Using the insertion of the effective action of the free real scalar superfield whose classical action looks like

$$S_v = -\frac{1}{16} \int d^8 z v D^\alpha \bar{D}^2 D_\alpha (\Box - M^2) v, \quad (4.259)$$

one can show that the one-loop effective action corresponding to the theory (4.258) can be expressed through the following Schwinger representation

$$\Gamma^{(1)} = \frac{i}{2}\text{Tr}\int \frac{ds}{s}\exp[is(\Box(\Box - M^2) + \frac{1}{4}\Psi\bar{D}^2 + \frac{1}{4}\bar{\Psi}D^2)]. \quad (4.260)$$

Since we restrict ourselves here to the Kählerian part of the effective action, we can express the one-loop effective action as

$$\Gamma^{(1)} = \frac{i}{2}\text{Tr}\int d^8z\int \frac{ds}{s}\exp[is(\frac{1}{4}\Psi\bar{D}^2 + \frac{1}{4}\bar{\Psi}D^2)]e^{is\Box(\Box-M^2)}\delta^8(z - z')|_{z=z'}. \quad (4.261)$$

The relevant terms from the operator $\exp(is(\frac{1}{4}\Psi\bar{D}^2 + \frac{1}{4}\bar{\Psi}D^2))$ again have the form (4.229), and, after Wick rotation $s = it$, the one-loop Kählerian effective action looks like

$$\Gamma_K^{(1)} = -i\int d^4\theta d^4x_1\int \frac{dt}{t}\frac{1}{\Box}[\cosh(t\sqrt{\Psi\bar{\Psi}\Box}) - 1]e^{-t\Box(\Box-M^2)}\delta^4(x_1 - x_2)|_{x_1=x_2}. \quad (4.262)$$

Expanding the hyperbolic cosine in series in $\Box$, after Wick rotation we find

$$\Gamma_K^{(1)} = \int d^4\theta d^4x_1\int \frac{dt}{t}\sum_{n=0}^{\infty}\frac{1}{(2n + 2)!}(t^2\Psi\bar{\Psi})^{n+1}\Box^n V(t; x_1, x_2)|_{x_1=x_2}. \quad (4.263)$$

Here the function $V(t; x_1, x_2)$ can be read off from the (4.249). As we already noted, this expression can be found in a closed form, for $M \neq 0$, only in the limit $M \to \infty$. It follows from (4.249) that

$$\Box^n V(t; x_1, x_2) = \frac{1}{t^n}(\frac{d}{d(M^2)})^n V(t; x_1, x_2), \quad (4.264)$$

so that, after taking $x_1 = x_2$,

$$\Box^n V(s; x_1, x_2)|_{x_1=x_2} = \frac{(-1)^n(n + 1)!}{16\pi^2(M^2t)^{n+2}}. \quad (4.265)$$

Taking all together, we find

$$\Gamma_K^{(1)} = \frac{1}{32\pi^2}\int d^8z\int \frac{dt}{M^2t^2}\sum_{n=0}^{\infty}\frac{(-1)^n(n + 1)!}{(2n + 2)!}\left(\frac{t\Psi\bar{\Psi}}{M^2}\right)^{n+1}. \quad (4.266)$$

This expression is similar to Eq. (4.176) obtained for the Wess-Zumino model. As a result, we have

$$K^{(1)} = \frac{1}{32\pi^2} \frac{\Psi\bar\Psi}{M^4} \sum_{n=0}^{\infty} \int_{\frac{\Psi\bar\Psi L^2}{M^4}}^{\infty} \frac{du}{u} \frac{(-1)^n u^n (n+1)!}{(2n+2)!}.$$ (4.267)

To avoid divergence of the integral, we introduced the cutoff L^2 at the lower limit. As $L^2 \to 0$, one obtains

$$K^{(1)} = -\frac{1}{32\pi^2} \frac{\Psi\bar\Psi}{M^4} \ln(\mu^2 L^2) - \frac{1}{32\pi^2} \frac{\Psi\bar\Psi}{M^4} (\ln \frac{\Psi\bar\Psi}{M^4\mu^2} - \xi).$$ (4.268)

Here ξ is some finite constant which can be absorbed into a redefinition of μ^2. This contribution is divergent but turns out to be suppressed in the large M limit. This divergence can be eliminated by adding a counterterm

$$K^{(1)}_{countr} = \frac{1}{32\pi^2} \frac{\Psi\bar\Psi}{M^4} \ln(\mu^2 L^2).$$ (4.269)

Thus, the renormalized Kählerian effective potential is

$$K^{(1)} = -\frac{1}{32\pi^2} \frac{\Psi\bar\Psi}{M^4} (\ln \frac{\Psi\bar\Psi}{M^4\mu^2} - \xi).$$ (4.270)

It is interesting also to proceed with the calculations in terms of the Feynman supergraphs, using the method similar to [75]. The one-loop effective action is described by the same supergraphs as at Fig. 4.11, with the external legs correspond to alternative Ψ and $\bar\Psi$. Their sum, calculated along the same lines as in the Sect. 4.8, after the Wick rotation is given by

$$\Gamma^{(1)} = -\frac{1}{2} \sum_{n=1}^{\infty} \frac{1}{n} \int d^4\theta \int \frac{d^4 k_E}{(2\pi)^4} (\Psi\bar\Psi \frac{D^2\bar D^2}{16 k_E^4 (k_E^2 + M^2)^2})^n \delta_{12}|_{\theta_1=\theta_2}.$$ (4.271)

The D-algebra transformations are simple, so we can easily sum the series and obtain

$$\Gamma^{(1)} = -\frac{1}{2} \int d^4\theta \int \frac{d^4 k_E}{(2\pi)^4} \frac{1}{k_E^2} \ln\left(1 + \frac{\Psi\bar\Psi}{k_E^2(k_E^2 + M^2)^2}\right).$$ (4.272)

This integral can be easiliy evaluated at $M = 0$, giving the result (4.240) which we have obtained above via the proper-time method. We note that this result can be obtained even without calculations. Indeed, by the symmetry reasons, the one-loop effective action can be only a function of $\Psi\bar\Psi$, and by the dimensional restrictions, the only answer for it is just the expression (4.240). However, if $M \neq 0$, the $\Gamma^{(1)}$ (4.272) can be evaluated in an easy form only in the limits $M \to 0$ or $M \to \infty$ as it has been done above. This methodology can be applied as well to the case of the presence of the higher-derivative gauge fields [98].

So, if $M \neq 0$, the situation is more complicated. Nevertheless, a possible solution in this case is presented in the paper [99]. Indeed, a typical structure of the one-loop Kählerian effective potential is given by the integral:

$$K^{(1)} = \frac{1}{2}\mu^\epsilon \int \frac{d^{4-\epsilon}k}{(2\pi)^{4-\epsilon}} \frac{1}{k^2} \log F(k^2), \qquad (4.273)$$

where $F(k^2)$ is a function of momenta whose form is determined by the quadratic action of quantum superfields. The simplest case is $F(k^2) = k^2 + g^2 \Phi\bar{\Phi}$ taking place in the Wess-Zumino model, see (4.182), and in supergauge theories, see [75, 100]. As we noted in previous sections, calculation of this integral is straightforward. In [99], it was noted that in higher-derivative theories with $2n$ derivatives, one can write $F(k^2) = (k^2 + A_1)(k^2 + A_2) \cdots (k^2 + A_n)$, where $A_1 \ldots A_n$ are the roots of $F(k^2)$ taken with opposite signs. In this case, the integral (4.273) can be easily taken, and the result will be a some function of roots $A_1 \ldots A_n$, an example is presented in [99]. However, while the integral over momenta for this representation of $F(k^2)$ is simple, the expressions for the roots $A_1, \ldots A_n$ are very complicated already at $n = 3$, implying in a very involved result of integrating (4.273), and certainly are more involved at higher values of n.

We considered the one-loop effective potential for two different versions of higher-derivative chiral superfield models. It turns out that, in the case when the mass term is purely chiral (a similar situation with the mass term takes place in the Wess-Zumino model), the theory is finite. At the same time, if the mass term arises in the general Lagrangian (that is the situation considered in [93]), the theory displays divergences in the limit of an infinite mass though being super-renormalizable. We note, however, that the equivalence of the higher-derivative theory of the chiral superfield and the theory without higher derivatives but with an extended number of chiral superfields described in [93] occurs only in the case when the mass term belongs to the general Lagrangian (that is, the second case considered in the section). Therefore, the presence of these divergences can be considered as a sign in favour of the equivalence established in [93]. Indeed, the expression (4.268), after relabelling $\frac{\Psi}{M^2} \to \Psi$, identically reproduces the one-loop Kählerian effective potential in the Wess-Zumino model, with $\Psi = m + \lambda\Phi$. We can therefore conclude that the higher-derivative theory (4.258), in the limit $M \to \infty$, yields the same quantum contribution to the effective potential as the Wess-Zumino model.

4.11 Supergauge Theories

This section is a brief review of results obtained for supergauge theories. Unfortunately, the restricted volume of this review does not allow to discuss all essential results of last years in this area hence we only give here main ones.

4.11.1 General Description of Supergauge Theories

The starting point of our consideration is the action of the $\mathcal{N} = 1$ SYM theory (cf. [66]):

$$S_{SYM} = \frac{1}{64g^2} \int d^6z \, \text{tr} \, W^\alpha W_\alpha, \tag{4.274}$$

where the field strength is

$$W_\alpha = -\bar{D}^2(e^{-gV} D_\alpha e^{gV}), \tag{4.275}$$

here the $V(z) = V^I(z)T^I$ is a real scalar Lie-algebra-valued superfield. We can expand the action (4.274) into power series in the coupling g. As a result we get

$$S = \frac{1}{16} \int d^8z \, \text{tr} \, (V D^\alpha \bar{D}^2 D_\alpha V + \ldots). \tag{4.276}$$

Here dots are for interaction terms. The action (4.274) is invariant under gauge transformations

$$e^{gV} \to e^{-ig\bar{\Lambda}} e^{gV} e^{ig\Lambda}. \tag{4.277}$$

where $\bar{D}_{\dot{\alpha}} \Lambda = 0$. The equivalent form of this transformation [33] is

$$\delta V = i L_{gV/2}(\Lambda + \bar{\Lambda} + \coth L_{gV/2}(\Lambda - \bar{\Lambda})). \tag{4.278}$$

Here $L_{gV} A = [gV, A]$ is a Lie derivative of an arbitrary superfield A. It is easy to see that strengths W_α, $\bar{W}_{\dot{\alpha}}$ transform covariantly under such transformations, while in the Abelian case they are invariant. The leading order of (4.278) at a small coupling is

$$\delta V = i(\Lambda - \bar{\Lambda}). \tag{4.279}$$

Since the theory is gauge invariant we must introduce gauge-fixing functions to perform the quantization. Their most natural form is

$$\chi(V) = -\frac{1}{4} \bar{D}^2 V + f(z) \tag{4.280}$$

$$\bar{\chi}(V) = -\frac{1}{4} D^2 V + \bar{f}(z).$$

Here $f(z)$ is an arbitrary chiral superfield (we note that in the Sect. 4.8 these gauge-fixing functions were used to calculate the one-loop effective action in the Wess-Zumino model). The variation of these gauge fixing functions under transformations (4.278) is

$$\delta \begin{pmatrix} \chi(V) \\ \bar{\chi}(V) \end{pmatrix} = \begin{pmatrix} 0 & -\frac{1}{4}\bar{D}^2 \\ -\frac{1}{4}D^2 & 0 \end{pmatrix} \begin{pmatrix} \delta V \\ \delta V \end{pmatrix}. \tag{4.281}$$

According to Faddeev-Popov approach we can introduce the ghost action defined as follows:

$$S_{GH} = i \int d^6z c' \delta \chi(V)|_{\Lambda \to c, \bar{\Lambda} \to \bar{c}} - i \int d^6\bar{z} \bar{c}' \delta \bar{\chi}(V)|_{\Lambda \to c, \bar{\Lambda} \to \bar{c}}, \tag{4.282}$$

i.e. parameters of the nonlinear gauge transformation Λ, $\bar{\Lambda}$ in this case are replaced by ghosts $c, \bar{c}$. Since the Λ is chiral, the c, c' are chiral ghosts, and the $\bar{c}, \bar{c}'$ are antichiral ones. As usual, ghosts are fermions.

Therefore the S_{GH} is

$$S_{GH} = i \int d^6z \, \text{tr} \, c' \frac{\delta \chi}{\delta V} \delta V - i \int d^6\bar{z} \, \text{tr} \, \bar{c}' \frac{\delta \bar{\chi}}{\delta V} \delta V, \tag{4.283}$$

where, on the base of (4.278), we can write

$$\delta V = iL_{gV/2}(c + \bar{c} + \coth L_{gV/2}(c - \bar{c})). \tag{4.284}$$

Hence, we arrive at the following action of ghosts:

$$S_{GH} = \int d^8z \, \text{tr} \, (\bar{c}' - c')L_{gV/2}(c + \bar{c} + \coth L_{gV/2}(c - \bar{c})). \tag{4.285}$$

Thus, the generating functional for this theory at zero sources, according to Faddeev-Popov approach, looks like

$$Z[J]|_{J=0} = \int DV D\{c\} e^{i(S_{SYM} + S_{GH})} \delta_+(\frac{1}{4}\bar{D}^2 V - f) \delta_-(\frac{1}{4}D^2 V - \bar{f}). \tag{4.286}$$

Here we use the notation $D\{c\} \equiv DcDc'D\bar{c}D\bar{c}'$ for an integral over all types of ghosts. As a next step of the Faddeev-Popov prescription, we can average over functions f and $\bar{f}$ with the weight

$$\exp(\frac{i}{\xi} \int d^8z(f\bar{f} + b\bar{b})), \tag{4.287}$$

where ξ is a some number (actually, it is a gauge parameter). The $b, \bar{b}$ are Nielsen-Kallosh ghosts (at the zero background, their contribution to effective action is a constant, but in the background-covariant formulation it is non-trivial). As a result, the (4.286) takes the form

$$Z[J]_{J=0} = \int DV D\{c\} e^{i(S_{SYM} + S_{GH} + S_{GF})}, \tag{4.288}$$

where

$$S_{GF} = \frac{1}{16\xi} \int d^8z \operatorname{tr}(\bar{D}^2 V)(D^2 V) \tag{4.289}$$

is the gauge-fixing action [33].

We introduce the total gauge-ghost action

$$S_{total} = S_{SYM} + S_{GF} + S_{GH} \tag{4.290}$$

and the corresponding generating functional

$$Z[J, \{\eta\}] = \int DV D\{c\} \exp(i(S_{total} + \int d^8z \operatorname{tr} JV + \int d^6z \operatorname{tr}(\eta'c' + \eta c) +$$
$$+ \int d^6\bar{z}(\bar{\eta}'\bar{c}' + \bar{\eta}\bar{c}))). \tag{4.291}$$

Here $\{\eta\}$ is the set of all sources: $\eta, \eta', \bar{\eta}, \bar{\eta}'$. To develop a diagram technique we must split the action (4.290) into a sum of the free (quadratic) part and vertices. It is easy to see [2, 33] that

$$e^{-gV} D_\alpha e^{gV} = g D_\alpha V - \frac{1}{2} g^2[V, D_\alpha V] + \frac{1}{6} g^3[V, [V, D_\alpha V]] + \dots. \tag{4.292}$$

Therefore (4.274) looks like

$$S_{SYM} = \int d^8z \operatorname{tr} \left(\frac{1}{16} V D^\alpha \bar{D}^2 D_\alpha V + \frac{1}{16} g(\bar{D}^2 D^\alpha V)[V, D_\alpha V] - \right.$$
$$\left. - \frac{1}{64} g^2[V, D^\alpha V]\bar{D}^2[V, D_\alpha V] - \frac{1}{48} g^2(\bar{D}^2 D^\alpha V)[V, [V, D_\alpha V]] + \dots \right). \tag{4.293}$$

And the ghost action is

$$S_{GH} = \int d^8z \operatorname{tr} \left(\bar{c}'c - \bar{c}c' + \frac{1}{2} g(\bar{c}' - c')[V, c + \bar{c}] + \frac{g^2}{12}(c' - \bar{c}')[V, [V, c - \bar{c}]] \right) +$$
$$+ \dots. \tag{4.294}$$

These expressions are sufficient for one- and two-loop calculations.

The quadratic action, with the added gauge-fixing term, is

$$S_0 = -\frac{1}{2} \int d^8z \operatorname{tr} V\left(-\frac{1}{8} D^\alpha \bar{D}^2 D_\alpha + \frac{1}{16\xi}\{D^2, \bar{D}^2\} \right) V + \int d^8z \operatorname{tr}(\bar{c}'c - \bar{c}c'). \tag{4.295}$$

The vertices can be read off from (4.293), (4.294). The propagators look like

$$\langle V^I(z_1)V^J(z_2)\rangle = i\delta^{IJ}\frac{1}{\Box}\Big(-\frac{1}{8\Box}D^\alpha\bar{D}^2D_\alpha + \xi\frac{\{D^2,\bar{D}^2\}}{16\Box}\Big)\delta^8(z_1-z_2); \quad (4.296)$$

$$\langle \bar{c}'^I(z_1)c^J(z_2)\rangle = \langle c'^I(z_1)\bar{c}^J(z_2)\rangle = -i\delta^{IJ}\frac{1}{\Box}\delta^8(z_1-z_2).$$

We note that ghosts are fermions, hence any ghost loop corresponds to minus sign. Then, D-factors are associated with vertices containing ghosts just by the same rule as with vertices containing any chiral superfields. We note that if we choose $\xi=1$ (the Feynman gauge) the propagator of gauge superfield takes the simplest form

$$\langle V^I(z_1)V^J(z_2)\rangle = i\delta^{IJ}\frac{1}{\Box}\delta^8(z_1-z_2). \tag{4.297}$$

Note that its sign is opposite to the sign of the propagator of a chiral superfield. This difference of signs plays an important role for some cancellations of divergences, see e.g. [23].

The action of a chiral superfield Φ coupled to a gauge one looks like

$$S = \int d^8z\,\bar{\Phi}_i(e^{gV})^i_j\Phi^j, \tag{4.298}$$

if $\Phi = \{\Phi_i\}$ is transformed under some representation of the gauge group (i.e. it is an isospinor), or

$$S = \int d^8z\,\mathrm{tr}(\bar{\Phi}e^{gV}\Phi e^{-gV}), \tag{4.299}$$

if $\Phi = \Phi^A T^A$ is Lie-algebra-valued. Note that under gauge transformations (4.277) the Φ is transformed as

$$\Phi \to e^{-ig\Lambda}\Phi \tag{4.300}$$

for an isospinor case and as

$$\Phi \to e^{-ig\Lambda}\Phi e^{ig\Lambda} \tag{4.301}$$

for a Lie-algebra-valued chiral superfield case. The Λ, $\bar{\Lambda}$ are Lie-algebra-valued parameters in both cases. The vertices can be easily obtained by expanding interaction parts of these actions into power series: for (4.298) we have

$$S_V = \int d^8z[\bar{\Phi}_i(e^{gV})^i_j\Phi^j - \Phi_i\bar{\Phi}^i] = \int d^8z\sum_{n=1}^{\infty}\frac{1}{n!}\bar{\Phi}_i((gV)^n)^i_j\Phi^j, \tag{4.302}$$

and for (4.299)

$$S_V = \text{tr} \int d^8z((\bar{\Phi}e^{gV}\Phi e^{-gV}) - \bar{\Phi}\Phi) = \text{tr} \int d^8z(g\bar{\Phi}[V,\Phi] + \frac{g^2}{2}\bar{\Phi}[V,[V,\Phi]] + \ldots). \quad (4.303)$$

The propagators of chiral superfields in this case are the standard ones given by (4.78), taken at $m = 0$ (we note that namely massless chiral superfields were considered in the Sect. 4.8 when the chiral effective potential in $\mathcal{N} = 1$ SYM theory coupled to a chiral matter was evaluated, and here we restrict out consideration to the massless case as well), and multiplied by delta symbols corresponding to algebraic indices.

The diagram technique derived now is very suitable for calculations in the sector of background (anti)chiral superfields only and for obtaining the divergences.

Let us consider an example. The $\mathcal{N} = 2$ SYM theory with a chiral matter is described by the action (see e.g. [2, 101]):

$$S = \frac{1}{64g^2} \int d^6z \, \text{tr} \, W^\alpha W_\alpha + \text{tr} \int d^8z \bar{\Phi}e^{gV}\Phi e^{-gV} +$$

$$+ \sum_{i=1}^{n} \left[ig(\int d^6z Q_i \Phi \tilde{Q}_i + h.c.) + \int d^8z \bar{\tilde{Q}}_i e^{-gV} \tilde{Q}_i + \int d^8z \bar{Q}_i e^{gV} Q_i \right].$$

$$(4.304)$$

Here Φ is the Lie-algebra-valued chiral superfield, and Q^i, $\tilde{Q}_i$ are chiral superfields transformed under mutually conjugated representations of Lie algebra. They are often called matter hypermultiplets. In the particular case $n = 1$, where we have only one superfield Q and one superfield $\tilde{Q}$, this action describes the $\mathcal{N} = 4$ SYM theory (the extended supersymmetry manifests itself through the so-called hidden supersymmetry transformations relating mutually the chiral and real superfields, see e.g. [23]).

We note that there is another equivalent formulation of the $\mathcal{N} = 4$ SYM theory characterized instead of Φ, Q and $\tilde{Q}$, by three Lie-algebra-valued chiral superfields Φ_i, with $i = 1 \ldots 3$, with the action (see [23]):

$$S = \frac{1}{64g^2} \int d^6z \, \text{tr} \, W^\alpha W_\alpha + \text{tr} \int d^8z \sum_{i=1}^{3} \bar{\Phi}_i e^{gV} \Phi_i e^{-gV} +$$

$$+ \frac{g}{3!}\epsilon_{ijk}\text{tr}\left[(\int d^6z \Phi_i \Phi_j \Phi_k + h.c.) \right]. \quad (4.305)$$

Let us consider the structure of one-loop divergences in the theory (4.304). For the sake of the simplicity we choose the Feynman gauge $\xi = 1$ in which the propagator is given by (4.297), therefore all tadpole diagrams given in [2] evidently vanish.

The contributions to the two-point function of the Φ field are given by Fig. 4.15.

Here and further, the thin line is for the propagator of the Lie-algebra valued chiral superfield Φ, the thick one—of hypermultiplets Q, $\tilde{Q}$, the wavy one—of the

Fig. 4.15 Contributions to the wave function renormalization of Φ

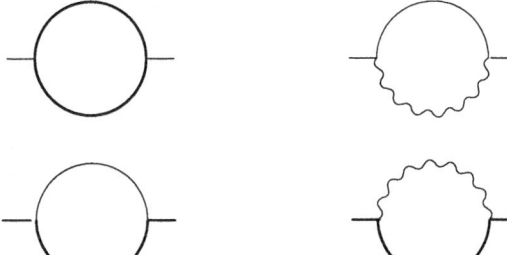

Fig. 4.16 Contributions to the wave function renormalization of $Q, \tilde{Q}$

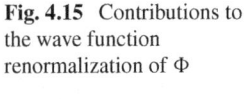

real superfield V, the dashed one—of ghosts. The D-factors in all these supergraphs are not shown, they are associated with vertices following standard Feynman rules.

One-loop divergent contributions from these supergraphs are respectively (see e.g. [101])

$$2 \sum_M \int \frac{d^4k}{(2\pi)^4} \frac{1}{k^2(k+p)^2} \Phi^A \bar{\Phi}^B \mathrm{tr}_M(T^A T^B) \tag{4.306}$$

and

$$-2 \int \frac{d^4k}{(2\pi)^4} \frac{1}{k^2(k+p)^2} \Phi^A \bar{\Phi}^B \mathrm{tr}_{ad}(T^A T^C T^D) \mathrm{tr}_{ad}(T^B T^C T^D). \tag{4.307}$$

Here tr_M denotes a trace in the representation under which the corresponding hypermultiplet is transformed, and $\sum_M$ is for the sum over hypermultiplets. Similarly, in (4.307), the traces are taken in the adjoint representation. The coefficient 2 is caused by the presence of two chiral hypermultiplets Q and $\tilde{Q}$ in the first term, and by two different contractions in the second term. We see that if $\sum_M \mathrm{tr}_M(T^A T^B) = \mathrm{tr}_{ad}(T^A T^C T^D) \mathrm{tr}_{ad}(T^B T^C T^D)$ there is no divergent contributions to wave function renormalization. In other words, the divergences in the matter sector are cancelled when the hypermultiplets are transformed under the specific representations of the gauge groups.

Contributions to the hypermultiplet two-point function are given by Fig. 4.16.

The one-loop divergent contributions from these supergraphs are respectively (cf. [101]):

$$\int \frac{d^4k}{(2\pi)^4} \frac{1}{k^2(k+p)^2} \bar{Q}_i Q_l (T^A)^{ij} (T^A)^{jl} \tag{4.308}$$

and

$$-\int \frac{d^4k}{(2\pi)^4} \frac{1}{k^2(k+p)^2} \bar{Q}_i Q_l (T^A)^{ij} (T^A)^{jl}. \tag{4.309}$$

Fig. 4.17 Contributions to the wave function renormalization of V

These corrections evidently cancel each other, hence the hypermultiplet wave function renormalization is trivial. In both these cases the cancellation is caused by the difference in signs of propagators of the gauge superfield and the chiral superfields (both Lie-algebra valued and the hypermultiplet ones). The same cancellation occurs for external $\tilde{Q}$ and $\bar{\tilde{Q}}$ legs.

Then, let us turn to the gauge sector of the theory. One-loop contributions to the wave function renormalization of the gauge superfield are represented by the supergraphs given at Fig. 4.17.

To make a brief description of the implications of the extended supersymmetry, we discuss now only the mutual cancellation of the quadratic divergences. Nevertheless, one can show [23], that the mutual cancellation of the logarithmic divergences within the background field method occurs for the same relations for the gauge group generators. So, the quadratically divergent contributions of these four supergraphs are respectively given by the following set of results (see e.g. [23, 101]):

$$\int \frac{d^4k}{(2\pi)^4} \frac{1}{k^2} V^A V^B \text{tr}_{ad}(T^A T^C T^D) \text{tr}_{ad}(T^B T^C T^D);$$

$$\int \frac{d^4k}{(2\pi)^4} \frac{1}{k^2} V^A V^B \text{tr}_{ad}(T^A T^C T^D) \text{tr}_{ad}(T^B T^C T^D);$$

$$2\sum_M \int \frac{d^4k}{(2\pi)^4} \frac{1}{k^2} V^A V^B \text{tr}_M(T^A T^B);$$

$$-4 \int \frac{d^4k}{(2\pi)^4} \frac{1}{k^2} V^A V^B \text{tr}_{ad}(T^A T^C T^D) \text{tr}_{ad}(T^B T^C T^D). \qquad (4.310)$$

Hence the same condition for cancellation of divergences for the two-point function of chiral superfields, that is,

$$\text{tr}_{ad}(T^A T^C T^D) \text{tr}_{ad}(T^B T^C T^D) = \text{tr}_M(T^A T^B), \qquad (4.311)$$

implies the cancellation of the quadratic divergences for the two-point function of the gauge superfield. As we have already mentioned, the same relation (4.311) results in the cancellation of the logarithmically divergent contributions. This condition is satisfied for an appropriate gauge group, in particular, $SU(N)$. The tadpole graphs (for the massless chiral superfield) give contributions proportional to the integral $\int \frac{d^4k}{(2\pi)^4} \frac{1}{k^2}$ which vanishes within the dimensional regularization, hence they are identically equal to zero. This mechanism explaining the vanishing of divergences is discussed, e.g. in [23, 102], where it is shown to be caused by the $\mathcal{N} = 2$ superconformal symmetry. The most important example of such theories is the $\mathcal{N} = 4$ SYM theory (4.305) which is equivalent to (4.304) with only one pair of hypermultiplets Q, $\tilde{Q}$ which are transformed under the adjoint representation of the Lie algebra. This theory is the first known example of a finite supersymmetric theory without higher derivatives (see e.g. [2, 23]).

4.11.2 Background Field Method

The approach to studying of supergauge theories described above is very useful for consideration of quantum corrections in the sector of chiral superfields Φ, Q, $\tilde{Q}$ only. To study contributions depending on gauge superfields we must develop a method allowing to preserve the manifest gauge invariance at any step, whereas earlier we have obtained contributions in terms of the superfield V which are in general case not gauge invariant. Therefore we should introduce an approach in which external lines are background strengths W_α, $\bar{W}_{\dot\alpha}$ and their *gauge covariant* derivatives. This method was developed in [66] (see also [106] and references therein), here we give its description.

The problem of calculation of the effective action in the SYM theory characterized by the action (4.274) is much more complicated than in scalar superfield theories. The main difficulties are the following ones. First, the nonpolynomiality of the action (4.274) implies in an infinite number of vertices which seems to result in infinite number of types of divergent quantum corrections (such a situation is treated in many cases as a non-renormalizability of the theory), second, it is easy to see that the usual background-quantum splitting $V \to V_0 + v$ where V_0 is a background field and v is a quantum field, cannot provide manifest gauge covariance of the quantum corrections. Really, because of the nonpolynomiality of W_α (4.275), to get a covariant quantum correction (which by definition must be expressed in terms of the strengths W_α, $\bar{W}_{\dot\alpha}$ which are the only objects transforming in a covariant way under the gauge transformations unlike of the superfield V itself) we need to summarize an infinite number of supergraphs with different numbers of external V_0 legs (to the best of our knowledge, this summation never has been performed). The background field method provides an efficient solution for these problems.

The starting point of this method is the *nonlinear* background-quantum splitting for the gauge superfield V defining the action (4.274) [66]:

$$e^{gV} \to e^{g\Omega} e^{gv} e^{g\bar{\Omega}}. \tag{4.312}$$

Here the v is a quantum field, the Ω, $\bar{\Omega}$ are the background superfields (they are not required to be chiral/antichiral ones, the only restriction is $e^{g\Omega}e^{g\bar{\Omega}} = e^{gV}$, with V is now a background gauge field, so, in principle one can have $\Omega = \bar{\Omega}$, or e.g. $\Omega = 0$). After such a background-quantum splitting (unfortunately, the complete proof of this statement is very tedious), the classical action (4.274) takes the form:

$$S = -\frac{1}{16g^2}\text{tr}\int d^8z(e^{-gv}\mathcal{D}^\alpha e^{gv})\bar{\mathcal{D}}^2(e^{-gv}\mathcal{D}_\alpha e^{gv}). \tag{4.313}$$

In this expression, describing the theory of the real scalar superfield v coupled to background superfields Ω, $\bar{\Omega}$, the $\mathcal{D}^\alpha$, $\bar{\mathcal{D}}^{\dot{\alpha}}$ are the background covariant derivatives defined by the expressions [23, 66]:

$$\mathcal{D}^\alpha = e^{-g\Omega}D^\alpha e^{g\Omega},$$
$$\bar{\mathcal{D}}^{\dot{\alpha}} = e^{g\bar{\Omega}}\bar{D}^{\dot{\alpha}}e^{-g\bar{\Omega}}. \tag{4.314}$$

Our further aim consists in the study of the action (4.313). To do it let us first describe the properties of the background covariant derivatives given by (4.314). As well as the covariant derivatives in usual differential geometry, the background covariant derivatives $\mathcal{D}^\alpha$, $\bar{\mathcal{D}}^{\dot{\alpha}}$ can be represented in the following "standard" form

$$\mathcal{D}^\alpha = D^\alpha - i\Gamma^\alpha, \quad \bar{\mathcal{D}}^{\dot{\alpha}} = \bar{D}^{\dot{\alpha}} - i\bar{\Gamma}^{\dot{\alpha}}, \tag{4.315}$$

where

$$\Gamma^\alpha = ie^{-g\Omega}(D^\alpha e^{g\Omega}), \quad \bar{\Gamma}^{\dot{\alpha}} = ie^{g\bar{\Omega}}(\bar{D}^{\dot{\alpha}}e^{-g\bar{\Omega}}) \tag{4.316}$$

are the superfield connections. Here, unlike (4.314), the derivatives act only to adjacent terms.

Let us study the (anti)commutation relations for the $\mathcal{D}^\alpha$, $\bar{\mathcal{D}}^{\dot{\alpha}}$. We start with imposing the following constraint

$$\mathcal{D}_{\alpha\dot{\alpha}} = -\frac{i}{2}\{\mathcal{D}_\alpha, \bar{\mathcal{D}}_{\dot{\alpha}}\}, \tag{4.317}$$

which represents itself as a background covariant analogue of the known anticommutation relation $\partial_{\alpha\dot{\alpha}} = -\frac{i}{2}\{D_\alpha, \bar{D}_{\dot{\alpha}}\}$. Then, it is easy to verify straightforwardly (e.g., at $\Omega = V$, and $\bar{\Omega} = 0$) the following definition of the background strength W_α (cf. [105]):

$$W_\alpha = [\bar{\mathcal{D}}^{\dot{\alpha}}, \{\bar{\mathcal{D}}_{\dot{\alpha}}, \mathcal{D}_\alpha\}] = 2i[\bar{\mathcal{D}}^{\dot{\alpha}}, \mathcal{D}_{\alpha\dot{\alpha}}]. \tag{4.318}$$

Really, after we substitute expressions (4.314) for the background-covariant derivatives to (4.318), and take into account that $e^{g\Omega}e^{g\bar{\Omega}} = e^{gV}$ in the case of absence of the quantum field v (cf. (4.312)), we get just the definition (4.275).

The relation (4.318) is crucial. It implies that the background-covariant space-time derivatives $\mathcal{D}_{\alpha\dot\alpha}$, unlike of the common covariant derivatives, have non-zero commutators with spinor background-covariant derivatives $\mathcal{D}_\alpha$, $\bar{\mathcal{D}}_{\dot\alpha}$, and, moreover, that the background strengths could arise during the D-algebra transformations. In particular, the identity (4.318) results in the following expressions (cf. [106]):

$$[\mathcal{D}_\alpha, \bar{\mathcal{D}}^2] = -W_\alpha + 4i\bar{\mathcal{D}}^{\dot\alpha}\mathcal{D}_{\alpha\dot\alpha} = W_\alpha + 4i\mathcal{D}_{\alpha\dot\alpha}\bar{\mathcal{D}}^{\dot\alpha}. \tag{4.319}$$

We also must define the (background) covariantly chiral superfields to describe the coupling of the gauge superfields to matter: the superfield Φ is referred as the (background) covariantly chiral one if it satisfies the condition $\bar{\mathcal{D}}_{\dot\alpha}\Phi = 0$. It is easy to see that the Φ is related to the usual chiral field Φ_0 as

$$\Phi = e^{g\bar\Omega}\Phi_0. \tag{4.320}$$

Really, condition of chirality $\bar{D}_{\dot\alpha}\Phi_0 = 0$ implies in

$$e^{g\bar\Omega}\bar{D}_{\dot\alpha}e^{-g\bar\Omega}e^{g\bar\Omega}\Phi_0 = 0. \tag{4.321}$$

Using the definitions (4.314), (4.320) we arrive just to the condition $\bar{\mathcal{D}}_{\dot\alpha}\Phi = 0$.

Now we can develop the perturbative approach for the theory with the action (4.313). First, we note that this theory possesses the symmetry with respect to the following gauge transformations:

$$e^{gv} \to e^{ig\bar\Lambda}e^{gv}e^{-ig\Lambda}, \tag{4.322}$$

where Λ is a *covariantly* chiral parameter, i.e. it satisfies the condition $\bar{\mathcal{D}}_{\dot\alpha}\Lambda = 0$ (similarly, $\mathcal{D}_\alpha\bar\Lambda = 0$). Therefore we need to introduce a gauge fixing. The most natural background covariant gauge fixing term looks like

$$S_{gf} = -\frac{1}{32}\mathrm{tr}\int d^8z\,v\{\mathcal{D}^2, \bar{\mathcal{D}}^2\}v, \tag{4.323}$$

which is a covariant generalization of the usual gauge fixing term in the Feynman gauge. Summarizing the (4.313) and (4.323), we get the following action of the quantum v field:

$$S_t = S + S_{gf} = -\frac{1}{2}\mathrm{tr}\int d^8z\,v\Box v + S_{int}, \tag{4.324}$$

where the S_{int} in the expression above is an interaction part including all W_α, $\bar{W}_{\dot\alpha}$ dependent terms, and $\Box = \mathcal{D}^m\mathcal{D}_m$ is the covariant d'Alembertian operator. In principle, one can fix other gauges (by introducing the factor ξ^{-1} in S_{gf}), however, even the problem of finding the propagator for v field appears to be very complicated for an arbitrary gauge, and up to now, nobody found this propagator in a closed form.

Hence the Feynman gauge is the most convenient one. The propagator of the v field in this gauge is

$$\langle v^I(z_1)v^J(z_2)\rangle = i\Box^{-1}\delta^{IJ}\delta^8(z_1 - z_2). \tag{4.325}$$

The interaction part of the total action of v includes both covariant generalizations of the usual vertices presented in (4.293) and new vertices involving the background strengths W_α, $\bar{W}_{\dot\alpha}$ manifestly (these expressions are analogues of those ones used in [66, 106], where, however, other conventions are adopted):

$$\begin{aligned}
S_{int} = \text{tr} \int d^8z\Big(&\frac{1}{2}v(W^\alpha\mathcal{D}_\alpha + \bar{W}_{\dot\alpha}\mathcal{D}^{\dot\alpha})v + \frac{1}{16}gv\{\mathcal{D}^\alpha v, \bar{D}^2\mathcal{D}_\alpha v\} - \\
&- \frac{1}{48}g[[\mathcal{D}^\alpha v, v], v]W_\alpha - \frac{1}{64}g^2[v, \mathcal{D}^\alpha v]\bar{D}^2[v, \mathcal{D}_\alpha v] - \\
&- \frac{1}{48}g^2\bar{D}^2\mathcal{D}^\alpha v[v, [v, \mathcal{D}_\alpha v]] - \frac{1}{192}g^2[[[\mathcal{D}^\alpha v, v], v], v]W_\alpha\Big) + \dots \tag{4.326}
\end{aligned}$$

We note that the terms proportional to W_α in the triple- and higher-order vertices in v arose from the anticommutation of gauge covariant derivatives in (4.313) between themselves, with using of the expression (4.318).

We also can prove the following important relation (cf. [66]):

$$\mathcal{D}^\alpha\bar{D}^2\mathcal{D}_\alpha - \frac{1}{2}\{\mathcal{D}^2, \bar{D}^2\} - W^\alpha\mathcal{D}_\alpha - \frac{1}{2}(\mathcal{D}^\alpha W_\alpha) = -8(\Box + W^\alpha\mathcal{D}_\alpha + \bar{W}_{\dot\alpha}\bar{D}^{\dot\alpha});$$

$$\mathcal{D}^2\bar{D}^2\mathcal{D}^2 = 16(\Box + \bar{W}_{\dot\alpha}\bar{D}^{\dot\alpha} + \frac{1}{2}(\bar{D}_{\dot\alpha}\bar{W}^{\dot\alpha}))\mathcal{D}^2 \equiv 16\Box_-\mathcal{D}^2;$$

$$\bar{D}^2\mathcal{D}^2\bar{D}^2 = 16(\Box + W^\alpha\mathcal{D}_\alpha + \frac{1}{2}(\mathcal{D}^\alpha W_\alpha))\bar{D}^2 \equiv 16\Box_-\bar{D}^2. \tag{4.327}$$

Because of the gauge symmetry, one needs to introduce ghosts. Since the form of the gauge transformations and the gauge fixing action is very similar to the usual superfield case evaluated at zero background fields in the previous subsection, with only difference consisting in the covariant chirality of matter and ghost fields instead of the usual one, the action of ghosts in this case is also analogous to the zero background case with only difference consisting in the *covariant* chirality of the ghosts c, c' and the *covariant* antichirality of the ghosts $\bar{c}$, $\bar{c}'$:

$$S_{gh} = \text{tr} \int d^8z(\bar{c}' - c')L_{gv/2}(c + \bar{c} + \text{cth}L_{gv/2}(c - \bar{c})), \tag{4.328}$$

which after expansion in the power series gives:

$$S_{gh} = \text{tr} \int d^8z(\bar{c}'c + c'\bar{c} + \frac{1}{2}(\bar{c}' - c')[v, c + \bar{c}] + \frac{1}{12}(c' - \bar{c}')[v, [v, c - \bar{c}]] + \dots). \tag{4.329}$$

The action of the matter after the background-quantum splitting of the gauge fields takes the form

$$S_m = \int d^8z \, \bar\Phi e^{gv} \Phi, \tag{4.330}$$

where Φ, $\bar\Phi$ are the *covariantly* chiral and antichiral superfields. The propagators of covariantly chiral superfields and ghosts are

$$\langle \bar\phi \phi \rangle = -i\Box_+^{-1}\delta^8(z_1 - z_2), \quad \langle \bar{c}'^I c^J \rangle = \langle \bar{c}^I c'^J \rangle = -i\delta^{IJ}\Box_+^{-1}\delta^8(z_1 - z_2), \tag{4.331}$$

with $\mathcal{D}^2$, $\bar{\mathcal{D}}^2$ factors are associated with the vertices in the same manner as D^2 and $\bar{D}^2$ in the case of usual chiral superfields. We note that such propagators can be expanded into power series in W_α, $\bar{W}_{\dot\alpha}$, see (4.327). The expressions (4.326), (4.329), (4.330), (4.325), (4.331) can be used for constructing the supergraphs. Some examples of the application of the method for supergraph calculations (in the pure SYM theory without matter) are presented in [106].

Now let us give a comparative characteristics for the two methods of superfield calculations—the background field method and the "usual" method considered in the previous subsection.

The crucial difference is the following one. In the framework of the "usual" superfield method, quadratic and linear divergences could arise for a supergraph with an arbitrary number of external gauge legs. Really, it is easy to show that the superficial degree of divergence for a "common" supergraph is

$$\omega = 2 - \frac{1}{2}N_D - E_\phi, \tag{4.332}$$

where N_D is a number of spinor supercovariant derivatives associated to the external legs, E_ϕ is a number of external chiral (antichiral) legs. We note that the quadratic and/or linear divergences are possible for any number of external v legs. At the same time, in the framework of the background field method, due to using the background-quantum splitting, the only external gauge legs are the background strengths and/or their covariant derivatives. The superficial degree of divergence in this case can be shown to have the form

$$\omega = 2 - \frac{3}{2}N_W - \frac{1}{2}N_D + \epsilon - E_\phi, \tag{4.333}$$

where N_W is a number of the background strength legs, $\epsilon = 1$ for the chiral (antichiral) contribution (the only its possible structure is $\int d^6z \, W^2$ or the conjugated one, $\int d^6\bar{z} \, \bar{W}^2$), otherwise $\epsilon = 0$; the N_D is the number of spinor supercovariant derivatives acting on external W_α, $\bar{W}_{\dot\alpha}$ legs (the derivatives presented in these W_α, $\bar{W}_{\dot\alpha}$'s must not be taken into account!). We see that within the framework of this approach only logarithmic overall divergences are possible, they arise for terms proportional to W^2, with the quadratic and linear subdivergences (which are important if we make a noncommutative generalization) can arise only in subgraphs which are not associated to the external W_α, $\bar{W}_{\dot\alpha}$ legs.

From formal viewpoint such a difference has the following origin. The superfield strength W_α by its construction contains three spinor derivatives, hence arising of any superfield strength leg in the framework of the usual superfield description decreases by three the number of the D-factors which could be converted to momenta (we note that the use of the background field method allows to sum automatically the infinite number of "usual" graphs and forbids an existence of supergraphs with a superficial quadratic or linear divergence). As a result, the convergence of a super-graph is improved. It is essential in the noncommutative field theory since it means that the only dangerous (quadratic and linear) infrared divergences could be generated by subgraphs of a given supergraph since each contribution to the effective action is in worst case only logarithmically divergent. In part, it means that there is no contradiction between the result of Bichl et al. [107] according to which the separate contributions to the one-loop two-point function of the gauge superfield in the $U(1)$ NC SYM theory possess quadratic divergences, and only their sum is free of dangerous UV/IR mixing, and the result of Zanon et al. [108] according to which all one-loop contributions to the effective action in the same theory are free of the dangerous UV/IR mixing. It is worth to notice that the calculations in the last paper were carried out in the framework of the background field method. We should mention also using of the background field method in the papers [108] devoted to study of the one-loop effective action in various noncommutative SYM theories. An important advantage of the background field method is that it allows to preserve the gauge covariance at all steps of calculations.

However, the background field method has one disadvantage—presence of non-trivial commutators of background covariant derivatives makes all calculations to be extremely difficult from the technical viewpoint even in the case of the absence of the external chiral matter fields.

4.11.3 *Proper-Time Method for the Supergauge Theories*

Within supergauge theories, there exists a powerful tool allowing for the application of the background field formalism in a manner based on the proper-time method. This method has been developed in [74]. To illustrate it, let us consider the following one-loop effective action of the gauge theories:

$$\Gamma^{(1)} = \frac{i}{2} \ln \det(\mathcal{D}^m \mathcal{D}_m + W^\alpha \mathcal{D}_\alpha + \bar{W}_{\dot\alpha} \bar{\mathcal{D}}^{\dot\alpha} + |\Phi|^2). \qquad (4.334)$$

Actually, this expression emerges when one couples the SYM theory, whose action is given by the sum of (4.324) and (4.326), to the external chiral matter. It has been discussed in [110]. Here we illustrate how this expression can be evaluated using the method developed in [74] where it was applied to obtain the contribution of the form

$W^2 f(\mathcal{D}W, \bar{\mathcal{D}}\bar{W})$, with $f(\mathcal{D}W, \bar{\mathcal{D}}\bar{W})$ is a some function of covariant derivatives of superfield strengths, reducing to a non-zero constant when these derivatives are equal to zero.

We start with the definition based on the well-known zeta function regularization procedure (see e.g. [109] for a review on this methodology):

$$\ln \det \Delta = -\zeta'(0), \qquad (4.335)$$

where the zeta function corresponding to the operator $\hat{K} = e^{t\Delta}$, with t is a proper time (see [36] for the general review on the proper time methodology) is defined as

$$\zeta(s) = \frac{1}{\Gamma(s)} \int_0^\infty dt\, t^{s-1} K(t), \qquad (4.336)$$

and $K(t)$ is a functional trace of $\hat{K}$. Explicitly, it looks like

$$K(t) = \int d^8z \lim_{z \to z'} e^{t\Delta} \delta^8(z - z'). \qquad (4.337)$$

Alternatively, one can use straightforwardly the Schwinger-De Witt representation which allows to write

$$\ln \det \Delta = \int_0^\infty \frac{dt}{t} \mathrm{tr} e^{t\Delta} \equiv \int_0^\infty \frac{dt}{t} K(t). \qquad (4.338)$$

Here we follow the methodology developed in [74] and use the notations adopted there. In our case, for the effective action given by (4.334), the $K(t)$ reads as

$$K(t) = \int d^8z \lim_{z \to z'} e^{t(\mathcal{D}^a \mathcal{D}_a + W^\alpha \mathcal{D}_\alpha + \bar{W}_{\dot{\alpha}} \bar{\mathcal{D}}^{\dot{\alpha}} + |\Phi|^2)} \delta^8(z - z'). \qquad (4.339)$$

Then, we use the Fourier representation of the complete delta function $\delta^8(z - z')$, both for its bosonic and fermionic parts:

$$\delta^8(z - z') = \int \frac{d^4k}{(2\pi)^4} e^{ik(x-x')} \int d^4\epsilon\, e^{i\epsilon^\alpha(\theta_\alpha - \theta'_\alpha)} e^{i\bar{\epsilon}_{\dot{\alpha}}(\bar{\theta}^{\dot{\alpha}} - \bar{\theta}'^{\dot{\alpha}})}. \qquad (4.340)$$

To simplify the calculations, we factorize out the chiral matter superfields (which are considered as constants within this calculation since we disregard all their derivatives), writing $K(t) = e^{t|\Phi|^2} \tilde{K}(t)$. Then, we introduce the key point of the methodology proposed in [74]: the $K(t)$ (4.339), being a kernel of the operator, is actually nothing other as the operator acting on the delta function, which further must be applied to some other function. Since the delta function is expanded into the Fourier series through (4.340), one can verify that when the operator whose kernel is given by (4.339) acts on an arbitrary function, the following objects will emerge:

$$X_m = \mathcal{D}_m + i k_m, \ X_\alpha = \mathcal{D}_\alpha + i \epsilon_\alpha, \ \bar{X}_{\dot\alpha} = \bar{\mathcal{D}}_{\dot\alpha} + i \bar{\epsilon}_{\dot\alpha}. \tag{4.341}$$

Indeed, for example, the purely spatial derivatives part in $K(t)$, acting on an arbitrary function $F(x')$ (in this case—the function of the bosonic coordinates only), will evidently produce the result

$$e^{t \mathcal{D}^a \mathcal{D}_a} e^{ik(x-x')} F(x')|_{x=x'} = e^{t X^m X_m} F(x). \tag{4.342}$$

Thus, acting of each derivative on the object $e^{ik(x-x')} F(x')$ or the similar one involving the Grassmannian delta function will extend this derivative with an additive term equal to the corresponding momentum multiplied by i. Repeating these arguments for spinor derivatives, we see that the kernel of the operator which would act on an arbitrary function of all superspace coordinates looks like

$$\tilde{K}(t) = \int d^8 z \int \frac{d^4 k}{(2\pi)^4} \int d^4 \epsilon \lim_{z \to z'} e^{t(X^a X_a + W^\alpha X_\alpha + \bar{W}_{\dot\alpha} \bar{X}^{\dot\alpha})}. \tag{4.343}$$

Now, let us differentiate this kernel with respect to the proper time t. It is straightforward to see that the derivative of $\tilde{K}(t)$ (4.343) looks like

$$\frac{d\tilde{K}(t)}{dt} = K_a^a + W^\alpha K_\alpha(t) + \bar{W}_{\dot\alpha} \bar{K}^{\dot\alpha}, \tag{4.344}$$

where

$$K_{A_1 \dots A_n} = \int \frac{d^4 k}{(2\pi)^4} \int d^4 \epsilon X_{A_1} \dots X_{A_n} e^{t\tilde{\Delta}} \tag{4.345}$$

have the role of n-th momenta of the generalized Gaussian (cf. [74]), and $\tilde{\Delta} = X^m X_m + W^\alpha X_\alpha + \bar{W}_{\dot\alpha} \bar{X}^{\dot\alpha}$. The expression (4.344) will be treated by us as the main equation of this study, similarly to the equation (4.130) for the Wess-Zumino model, aimed for calculating the corresponding heat kernel.

Then, following [74], we use some identities representing themselves as integrals over the whole space of the momenta from total derivatives with respect to momenta $k_a, \epsilon_\alpha, \bar{\epsilon}_{\dot\alpha}$ defined in (4.340):

$$\int \frac{d^4 k}{(2\pi)^4} \int d^4 \epsilon \frac{\partial}{\partial \epsilon_\alpha} (X_{A_1} \dots X_{A_n} e^{t\tilde{\Delta}}) = 0;$$

$$\int \frac{d^4 k}{(2\pi)^4} \int d^4 \epsilon \frac{\partial}{\partial \bar{\epsilon}_{\dot\alpha}} (X_{A_1} \dots X_{A_n} e^{t\tilde{\Delta}}) = 0;$$

$$\int \frac{d^4 k}{(2\pi)^4} \int d^4 \epsilon \frac{\partial}{\partial k_a} (X_{A_1} \dots X_{A_n} e^{t\tilde{\Delta}}) = 0. \tag{4.346}$$

It follows from the structure of (4.343) that we must take into account only terms with one X_α (or $X_{\dot\alpha}$) and no more than two X_a. So, the identities (4.346) imply in the need to consider the following expression:

$$\frac{\partial}{\partial\epsilon_\alpha}e^{t\tilde\Delta} = -it\sum_{n=0}^{\infty}\frac{t^n}{(n+1)!}ad^{(n)}(\tilde\Delta)(W^\alpha)e^{t\tilde\Delta}, \tag{4.347}$$

where $ad(\tilde\Delta)(W_\beta) = [\tilde\Delta, W_\beta]$, $ad^{(2)}(\tilde\Delta)(W_\beta) = [\tilde\Delta, [\tilde\Delta, W_\beta]]$, etc. The expression for the derivative with respect to $\bar\epsilon_{\dot\alpha}$ is a straightforward analogue of this one, and that one involving the derivative with respect to k_a will be introduced further.

If we want to restrict ourselves to expressions involving, at most, first derivatives of any superfield strengths (note that just these expressions, being projected to the components, give different degrees of the stress tensor F_{ab}, whereas the higher derivatives of W_α, $\bar W_{\dot\alpha}$ imply in the terms involving the derivatives of F_{ab} which are irrelevant for our purposes), we must consider the only nontrivial commutators:

$$[X_a, X_b] = -\frac{1}{2}(\bar D\bar\sigma_{ab}\bar W - D\sigma_{ab}W) \equiv -\frac{1}{2}(\bar M_{ab} - M_{ab}); \quad [X_a, X_\alpha] = i(\sigma_a)_{\alpha\dot\alpha}\bar W^{\dot\alpha};$$

$$\{X_\alpha, W_\beta\} = (D_\alpha W_\beta) = N_{\alpha\beta}; \quad \{\bar X_{\dot\alpha}, \bar W_{\dot\beta}\} = (\bar D_{\dot\alpha}\bar W_{\dot\beta}) = \bar N_{\dot\alpha\dot\beta}. \tag{4.348}$$

Now, it is crucial that the background fields belong to the Abelian subalgebra of the gauge algebra (cf. [110]). So, one can write $ad(\tilde\Delta)(W_\beta) = [\tilde\Delta, W_\beta] = W^\alpha N_{\alpha\beta}$. Repeating the calculation of the commutator n times, we find that

$$ad^{(n)}(\tilde\Delta)(W_\beta) = W^\alpha(N^n)_{\alpha\beta}. \tag{4.349}$$

Then, we make use of the identities (4.346). The first one looks like

$$0 = \int\frac{d^4k}{(2\pi)^4}\frac{\partial}{\partial\epsilon_\beta}(X_\alpha e^{t\tilde\Delta}) = i\delta_\alpha^\beta\tilde K(t) - \int\frac{d^4k}{(2\pi)^4}X_\alpha\frac{\partial}{\partial\epsilon_\beta}e^{t\tilde\Delta}. \tag{4.350}$$

We employ the expression (4.347) to find the derivative with respect to ϵ_β, and, afterwards (4.349), to find n-th adjoint of W_α. As a consequence, we find that

$$\frac{\partial}{\partial\epsilon_\beta}e^{t\tilde\Delta} = -it\sum_{n=0}^{\infty}\frac{t^n}{(n+1)!}W^\gamma(N^n)_\gamma{}^\beta = -iW^\gamma(\frac{e^{tN}-1}{N})_\gamma{}^\beta. \tag{4.351}$$

We substitute these expressions to (4.350). Then, it remains to carry out the anticommutation between X_α and W^γ, that is, $\{X_\alpha, W^\gamma\} = N_\alpha{}^\gamma$. We arrive at

$$0 = \delta_\alpha^\beta\tilde K(t) + N_\alpha{}^\gamma(\frac{e^{tN}-1}{N})_\gamma{}^\beta\tilde K(t) - W^\gamma(\frac{e^{tN}-1}{N})_\gamma{}^\beta K_\alpha(t). \tag{4.352}$$

Moving the term with $W^\gamma K_\alpha(t)$ to the left-hand side of the expression, multiplying this equation by the matrix inverse to $(\frac{e^{tN}-1}{N})$, and calculating the trace (with imposing the restriction $N^\alpha_\alpha = 0$), we arrive at

$$W^\alpha K_\alpha(t) = \text{tr}(\frac{N}{e^{tN}-1})\tilde{K}(t). \tag{4.353}$$

Proceeding in a similar way for the identity conjugated to (4.350), we find

$$\bar{W}_{\dot{\alpha}}\bar{K}^{\dot{\alpha}}(t) = \text{tr}(\frac{\bar{N}}{e^{t\bar{N}}-1})\tilde{K}(t). \tag{4.354}$$

We note that this term was absent in [74] where only the contribution dependent on W_α, but not on $\bar{W}_{\dot{\alpha}}$ was considered.

Finally, one can identically repeat the calculation performed in [74], to obtain the $K_{ab}(t)$. One starts with the identity

$$0 = \int \frac{d^4k}{(2\pi)^4}\frac{\partial}{\partial k_b}(X_a e^{t\tilde{\Delta}}) = i\delta_{ab}\tilde{K}(t) + \int \frac{d^4k}{(2\pi)^4}X_a\frac{\partial}{\partial k_b}e^{t\tilde{\Delta}}, \tag{4.355}$$

and, similarly to the calculations above, one finds

$$\frac{\partial}{\partial k_b}e^{t\tilde{\Delta}} = \sum_{n=0}^{\infty}\frac{t^n}{(n+1)!}ad^{(n)}(X\cdot X)(X_b) = 2it B_{bc}(t)X_c, \tag{4.356}$$

where

$$B_{bc} = \left(\frac{e^{-t(\bar{M}-M)}-1}{-t(\bar{M}-M)}\right)_{bc}. \tag{4.357}$$

Therefore, restoring the K_{ac} with use of its definitions following from (4.345), one finds that the identity (4.355) leads to

$$0 = i\delta_{ab}\tilde{K}(t) + 2it B_{bc}(t)K_{ac}(t) \tag{4.358}$$

(the term involving $K_a(t)$ will be irrelevant just as in [74]), so, one has

$$K_{ab}(t) = -\frac{1}{2t}(B^{-1})_{ba}(t)\tilde{K}(t) = \frac{1}{2}\left(\frac{\bar{M}-M}{e^{-t(\bar{M}-M)}-1}\right)_{ba}\tilde{K}(t). \tag{4.359}$$

For the sake of simplicity, we suggested within these calculations that $N^\alpha_\alpha = 0$, as well as $N^{\dot{\alpha}}_{\dot{\alpha}} = 0$. These identities can be imposed since these expressions do not contribute to the degrees of freedom of the stress tensor F_{ab}; actually, for the Abelian background superfield they are just equivalent to the Bianchi identities $\{\mathcal{D}^\alpha, W_\alpha\} =$

0, $\{\bar{D}_{\dot{\alpha}}, \bar{W}^{\dot{\alpha}}\} = 0$. Substituting the expressions (4.353), (4.354), (4.359) to (4.344), we arrive at

$$\frac{d\tilde{K}(t)}{dt} = \left[\frac{1}{2}\left(\frac{\bar{M} - M}{e^{-t(\bar{M}-M)} - 1}\right)^a_a + \text{tr}\left(\frac{N}{e^{tN} - 1}\right) + \text{tr}\left(\frac{\bar{N}}{e^{t\bar{N}} - 1}\right)\right]\tilde{K}(t).$$

(4.360)

The solution of this equation is

$$\tilde{K}(t) = \frac{W^2 \bar{W}^2}{16\pi^2}\det\left(\frac{e^{tN} - 1}{N}\right)\det\left(\frac{e^{t\bar{N}} - 1}{\bar{N}}\right)\det\left(\frac{1 - e^{-t(\bar{M}-M)}}{M - \bar{M}}\right)^{-1/2}.$$

(4.361)

The overall constant factor is fixed from the fact that, similarly to [74], the solution of this equation at $M = \bar{M} = N = 0$ must be $K(t) = \frac{W^2 \bar{W}^2}{16\pi^2}$, that is, the expression which yields the well-known result for the four-point function of W_α and $\bar{W}_{\dot{\alpha}}$ but not on their derivatives [23, 111]. On the base of this kernel, one can write down the following one-loop effective action:

$$\Gamma^{(1)} = \int d^8 z \int \frac{dt}{t} e^{-t|\Phi|^2} \frac{W^2 \bar{W}^2}{16\pi^2}\det\left(\frac{e^{tN} - 1}{N}\right)\det\left(\frac{e^{t\bar{N}} - 1}{\bar{N}}\right) \times$$

$$\times \det\left(\frac{1 - e^{-t(\bar{M}-M)}}{M - \bar{M}}\right)^{-1/2}.$$

(4.362)

We note that due to the identity [74]:

$$\det\left(\frac{1 - e^{-2tF}}{F}\right)^{-1/2} = \frac{1}{4t^2}\det\left(\frac{tF}{\sinh tF}\right)^{1/2},$$

(4.363)

this expression can be rewritten in an alternative form:

$$\Gamma^{(1)} = \int d^8 z \int \frac{dt}{t^3} e^{-t|\Phi|^2} \frac{W^2 \bar{W}^2}{16\pi^2}\det\left(\frac{e^{tN} - 1}{N}\right)\det\left(\frac{e^{t\bar{N}} - 1}{\bar{N}}\right) \times$$

$$\times \det\left(\frac{t(\bar{M} - M)}{\sinh t(\bar{M} - M)}\right)^{1/2}.$$

(4.364)

This is just the result obtained in [110] for the $SU(2)$ gauge group spontaneously broken to $U(1)$. We note that if one will suggest that the derivatives of the strengths are zero, i.e. $M = \bar{M} = N = 0$, one will have $\det(\frac{e^{tN}-1}{N})|_{N=0} = t^2$ (remind that N is 2×2 matrix since the spinor indices take values 1 and 2), one recovers the well-known result [23, 111]:

$$\Gamma^{(1)} = \frac{1}{16\pi^2} \int d^8z \, \frac{W^2 \bar{W}^2}{(\Phi \bar{\Phi})^2}. \tag{4.365}$$

Unlike [74], we obtained the results depending both on W_α and $\bar{W}_{\dot\alpha}$.

This study can be easily generalized for the $SU(n)$ gauge group broken to its maximal Abelian subgroup (so-called Abelian torus) $U(1)^{n-1}$. Indeed, following [111], we can write the one-loop effective action in this case as

$$\Gamma^{(1)} = \frac{i}{2} \sum_{k<l} \ln \det(-\mathcal{D}^a \mathcal{D}_a - (W_k^\alpha - W_l^\alpha)\mathcal{D}_\alpha - (\bar{W}_{\dot\alpha k} - \bar{W}_{\dot\alpha l})\bar{\mathcal{D}}^{\dot\alpha} - |\Phi_k - \Phi_l|^2). \tag{4.366}$$

Here $W_k^\alpha - W_l^\alpha$ etc. are the superfield roots of the $su(n)$ algebra (see the detailed discussion of the algebra roots for $su(n)$ and a derivation of this expression in [111]). Explicitly repeating the calculation above for (4.366), we arrive at

$$\Gamma^{(1)} = \sum_{k<l} \int d^8z \int \frac{dt}{t} e^{-t|\Phi_{kl}|^2} \frac{W_{kl}^2 \bar{W}_{kl}^2}{16\pi^2} \det\left(\frac{e^{tN_{kl}} - 1}{N_{kl}}\right) \det\left(\frac{e^{t\bar{N}_{kl}} - 1}{\bar{N}_{kl}}\right) \times$$
$$\times \det\left(\frac{1 - e^{-t(\bar{M}_{kl} - M_{kl})}}{M_{kl} - \bar{M}_{kl}}\right)^{-1/2}, \tag{4.367}$$

where $W_{kl}^\alpha = W_k^\alpha - W_l^\alpha$, $\Phi_{kl} = \Phi_k - \Phi_l$ etc. We close this section with the conclusion that the one-loop effective action of the SYM theory, in the approximation of constant background fields F_{ab} and ϕ (that is, the principal components of W_α and Φ), has been successfully calculated for the gauge group $SU(n)$, with an arbitrary n. In principle, the calculation for other gauge groups will not essentially differ.

4.12　Conclusions

We described the superfield formalism in the four-dimensional space-time. Our main conclusions are, first, that the superfield approach allows for a very compact manner of perturbative calculations, second, that it allows to take into account the famous "miraculous cancellations" automatically, simplifying this the analysis of possible divergences.

Within this chapter, we considered various superfield models describing two most used supersymmetric multiplets, the chiral one and the real one. The importance of these multiplets is motivated by the fact that, first, the real Lie-algebra valued superfield allows for generalizing the Yang-Mills theory to a superspace introducing thus a class of SYM models and hence implying a possibility for constructing superfield models of all fundamental interactions, except of the gravitational one, second, the chiral superfield is treated as one of the most natural representations of supersymmetric matter. One more reason of the interest to SYM theories is caused by the fact that namely the $\mathcal{N} = 4$ SYM theory is the first known explicitly finite four-dimensional theory without higher derivatives. It is interesting to note that this finiteness can be apparently extended to the noncommutative generalization of this theory, see [104]. Moreover, we explicitly demonstrated that our methodology can be applied not only to usual Wess-Zumino and supergauge theories but can be also extended to their nontrivial generalizations including non-polynomial and higher-derivative ones, which play important role for constructing various low-energy effective models (as we already argued, such effective models can arise, for example, when some of superfields, for example heavy ones, are integrated out). It is interesting to note that the superfield approach is very efficient for performing quantum calculations even in nonlocal supersymmetric theories, i.e. those ones involving infinite orders in derivatives. The examples of such studies are presented in [112] (see also references therein).

At the same time, the natural question is—whether perturbative treating of superfield models based on other multiplets is possible and can be realized? An important example of such a multiplet is the spinor chiral superfield representing the so-called tensor multiplet defined in [23], whose importance is motivated by the fact that its component content includes the Kalb-Ramond antisymmetric tensor field, therefore it can play a fundamental role within study of effective theories arising in the string context. An important property of this superfield consists in the fact that its free Lagrangian possesses the gauge symmetry. First perturbative studies of field theory models involving this superfield have been performed in [113], where the one-loop effective action in a theory involving a spinor chiral superfield and a usual real gauge scalar superfield coupled to a chiral matter has been calculated.

And, besides of considering these superfields, studies of the supergravity multiplet certainly have a fundamental role. A detailed discussion of the supergravity is presented in [23]; here, because of restricted volume of our review, we will not discuss it.

Chapter 5
Supersymmetry Breaking

Throughout these lecture notes, we have discussed the superfield methodology for studying supersymmetric field theories. However, it is known that, actually the supersymmetry is broken at observed scales of energy. There are two ways to describe the breaking of any symmetry including the supersymmetry, those are—explicit and spontaneous breaking. The detailed discussions of the supersymmetry breaking are presented in famous review papers, such as [114, 115], and, of course, in classical supersymmetry textbooks like [23, 32, 43]. Here we do not intend to give a detailed discussion of the supersymmetry breaking, following a more modest aim—to present a brief description of some ways to describe these phenomena in terms of the superfield methodology. Here we concentrate on the theories defined in the four-dimensional space-time.

5.1 Explicit Supersymmetry Breaking

To break the supersymmetry, continuing within the framework of the superfield approach, we suggest that the classical action of the theory involves a small additive term whose presence breaks the supersymmetry. Proceeding on the base of the superfield description, at least formally, we can introduce such an additive term through an introduction of a special extra "superfield" possessing a broken component structure, with a corresponding non-trivial component of this superfield is a constant. Actually such a superfield, called a *spurion*, represents itself as a some generalization of a coupling constant (actually, this is a soft supersymmetry breaking if the divergences continue to be logarithmic; we also note that additive supersymmetry-breaking terms are small). There is a natural restriction on the structure of such terms, indeed, it is easy to see that only logarithmic divergences can emerge within quantum corrections in Wess-Zumino and SYM theories: while for the Wess-Zumino model it follows

A. Petrov, *Quantum Superfield Supersymmetry*, Fundamental Theories of Physics 202, https://doi.org/10.1007/978-3-030-68136-4_5

directly from the study of the superficial degree of divergence, for the SYM theory
the higher-order divergences can be shown to be forbidden by the gauge symmetry.
Therefore, we suppose the spurion terms to be in the form which does not introduce
quadratic divergences. In particular, it is necessary to require the constant defining
the spurion to have a non-negative mass dimension.

The key idea of the supersymmetry breaking is the following one [114]: it is easy
to see that the above-mentioned constant cannot be the lower ($\theta, \bar{\theta}$-independent)
component of the superfield since in this case the variation of the superfield under
the supersymmetry transformations vanishes, hence the supersymmetry will not be
broken. Therefore, there are three possible monomial forms of the spurion (we note
that the constant fields μ, $\bar{\mu}$, ν must be scalar, to maintain Lorentz invariance), cf.
[23]:

$$\chi = \mu^2 \theta^2, \quad \bar{\chi} = \bar{\mu}^2 \bar{\theta}^2, \quad U = \nu^2 \theta^2 \bar{\theta}^2. \tag{5.1}$$

Two first types of spurions can emerge in chiral and antichiral sectors of the action
respectively, and the third one—only in a general sector. The corresponding additional terms in the action of the Wess-Zumino model (and models including it as an
ingredient), respectively, look like

$$\Delta S_1 = \int d^6 z \chi \Phi^2 + h.c. = \int d^4 x (\mu^2 \varphi^2 + h.c.),$$

$$\Delta S_2 = \int d^8 z U \Phi \bar{\Phi} = \int d^4 x \nu^2 \varphi \bar{\varphi}, \tag{5.2}$$

where φ is a scalar component of the chiral superfield Φ. A brief inspection of the
component structure of the resulting theory shows that these additive terms, although
they do not generate new types of counterterms, destroy the equality of masses of
bosonic and fermionic fields which is known to be characteristic for supersymmetric
theories, thus, the supersymmetry in these cases is broken. We note that the mass
dimension of μ^2 is 2, and of ν^2 is zero, i.e. in both cases it is non-negative, and the
renormalizability of the theory is not jeopardized.

It is easy to verify that if the mass dimension of the spurion superfield is non-
negative, no new divergent terms can emerge. Indeed, in this case only the vertices
with degrees of superfields and/or derivatives no higher than those ones given in
the initial action of the theory will arise, therefore there is no possibility for non-
renormalizable interactions.

However, in principle, there are non-renormalizable spurion couplings, like e.g.
$\int d^8 z U D^\alpha \Phi D_\alpha \Phi$ [23]. In this case the spurion has a negative dimension. From the
viewpoint of the Feynman supergraphs, each spurion vertex in this case involves two
extra spinor derivatives which increases a superficial degree of divergence of the
corresponding supergraph. Such a manner of supersymmetry breaking is evidently
not soft.

We close this section with a mentioning that the soft supersymmetry breaking has wide phenomenological and cosmological applications, for example, in the context of the dark matter problem, see e.g. [116].

5.2 Spontaneous Supersymmetry Breaking

Now, let us suggest that the classical action of the theory is supersymmetric and formulated in terms of superfields. At the same time, we suppose that the vacuum is not supersymmetric, so, the supersymmetry is broken in a spontaneous manner. This way of supersymmetry breaking seems to be more delicate being a principal subject of many studies.

The key observation which gave rise to many discussions of the spontaneous supersymmetry breaking is the following one (it is explained, e.g., in [23, 32] and many other textbooks): in any supersymmetric theory, the Hamiltonian, that is, the energy operator, can be written in terms of the supersymmetry generators as

$$\hat{H} \equiv \hat{P}^0 = \frac{1}{4}(\sigma^0)^{\alpha\dot{\beta}}\{Q_\alpha, \bar{Q}_{\dot{\beta}}\} = \frac{1}{4}(\{Q_1, \bar{Q}_{\dot{1}}\} + \{Q_2, \bar{Q}_{\dot{2}}\}), \qquad (5.3)$$

so, the Hamiltonian of supersymmetric theories is positively defined, with the supersymmetry generators play the role of the creation and annihilation operators. As a result, acting of the annihilation operator on the vacuum state $|0>$ should yield zero, $\bar{Q}_{\dot{\alpha}}|0>= 0$. Similarly, one has $Q_\alpha|0>= 0$. Therefore, the variation of the vacuum under the supersymmetry transformations is zero, $(\epsilon^\alpha Q_\alpha + \bar{\epsilon}_{\dot{\alpha}}\bar{Q}^\alpha)|0>= 0$, i.e. vacuum is invariant under supersymmetry transformations. Using (5.3), one finds that in this case $\hat{H}|0>= 0$. It is clear that a minimum of the Hamiltonian of a field theory is a minimum of its potential since the kinetic energy is evidently non-negative. Therefore, we have a natural criterium: if the vacuum (that is, the lowest value of the potential) of a supersymmetric theory is zero, the supersymmetry is not spontaneously broken.

Now, let us see situations where the supersymmetry is spontaneously broken. First, one of the simplest examples of the models involving the spontaneous supersymmetry breaking is the super-QED extended by the additive, gauge invariant Fayet-Iliopoulos term [117]:

$$S_{FI} = -\xi \int d^8z V. \qquad (5.4)$$

This term is linear in the superfield. Moreover, in components it has the simple form $\xi \int d^4x \mathcal{D}(x)$, where $\mathcal{D}$ is higher component of the gauge superfield (4.24). It is clear that this term breaks the parity $V \to -V$ (and, consequently, $\mathcal{D} \to -\mathcal{D}$). Since this field enters the action of the super-QED (cf. (4.53)) only through the term

$$S_D = \int d^4x (\frac{1}{2}\mathcal{D}^2 - \xi\mathcal{D}), \tag{5.5}$$

it is clear that the equations of motion for $\mathcal{D}$, besides of the usual solution $\mathcal{D} = 0$, yield also the solution $\mathcal{D} = \xi$ which does not possess the symmetry $\mathcal{D} \to -\mathcal{D}$, so, this symmetry is broken. The supersymmetry is also evidently broken since the supersymmetry transformations near this vacuum, for some components of the superfield V (in particular, for the vector field A_a) will be ξ-independent, but for some (in particular, for the photino λ_α)—ξ-dependent. Introducing of interaction of the gauge superfield with the chiral matter will not essentially modify the situation [23].

Other important example of the models displaying the spontaneous supersymmetry breaking are the O'Raifeartaigh models [118]. In models of this class, one has a set of chiral superfields whose kinetic terms are the standard ones, but the potential is more sophisticated (although renormalizable). The simplest example of such models is given by the action [118, 119]:

$$S = \int d^8z(\bar{X}X + \bar{\phi}_1\phi_1 + \bar{\phi}_2\phi_2) + \left[\int d^6z(m\phi_1\phi_2 + hX\phi_1^2 + fX) + h.c.\right], \tag{5.6}$$

where X and $\phi_{1,2}$ are chiral superfields. If one would obtain the equations of motion for this theory and then put $D^2\phi_{1,2} \simeq 0$, $D^2X \simeq 0$, in order to consider only slowly varying superfields (that is, to restrict oneself to considering only the Kählerian potential), the following system of equations arises:

$$f + h\phi_1^2 = 0;$$
$$m\phi_2 + hX\phi_1 = 0;$$
$$m\phi_1 = 0. \tag{5.7}$$

Such a system is evidently inconsistent, therefore, the vacuum for this theory simply does not exist, and the supersymmetry is broken. It was shown in [119] that a whole class of superfield theories possesses a similar behavior.

We close this section with recommending of the brilliant review [115] for a further reading on this subject.

Chapter 6
Conclusions

In this book, we considered the superfield method in supersymmetric field theories, both in three- and four-dimensional space-times. This method allows to preserve the manifest supersymmetry at any step of calculations, and the calculations within it turn out to be much more compact than within the framework of the component approach.

We studied several examples of superfield theories and presented in details quantum calculations for these models. In three dimensions, our examples were a scalar superfield theory and supersymmetric gauge theories. In four dimensions, we considered the Wess-Zumino model, the general chiral superfield model, the higher-derivative chiral superfield model and the $\mathcal{N} = 1$ SYM theory with chiral matter. In all these theories we developed the supergraph technique, studied a general form of a superfield effective action and calculated low-energy leading contributions to the effective actions of these theories in one-, and in certain cases, two-loop approximations. It is natural to expect that the development of superfield quantum calculations in other superfield models formulated in terms of $\mathcal{N} = 1$ superfields is in principle no more difficult. We also discussed the background field method in supergauge theories.

Let us briefly discuss other applications and generalizations of the superfield approach in the quantum field theory. In the last years the following most important ways of applications of superfield supersymmetry were developed.

1. *Studying of theories with extended supersymmetry.* It is known that theories with extended supersymmetry possess better renormalization properties, e.g. while the $\mathcal{N} = 1$ SYM theory is renormalizable, the $\mathcal{N} = 4$ SYM theory is finite. The most important examples of theories possessing extended supersymmetry are $\mathcal{N} = 2$ and $\mathcal{N} = 4$ SYM theories. During last years numerous results in studying of these theories were obtained (see e.g. [110, 111, 120] and references therein). It is natural that the most adequate method for considering such theories must display manifest $\mathcal{N} = 2$ supersymmetry. Such a method is a harmonic superspace

A. Petrov, *Quantum Superfield Supersymmetry*, Fundamental Theories of Physics 202, https://doi.org/10.1007/978-3-030-68136-4_6

approach developed in [19, 20]. This method is based on defining superfields as functions of bosonic space-time coordinates x^a, two sets of Grassmannian coordinates $\theta^{i\alpha}$, $\bar{\theta}^{i\dot{\alpha}}$ with $i = 1, 2$ and spherical harmonics $u^{\pm i}$. Introducing of the analytic superfield [19] allows to develop formulation of various theories in terms of unconstrained $\mathcal{N} = 2$ superfields and to avoid arising of component fields with higher spins. The $\mathcal{N} = 2$ and $\mathcal{N} = 4$ SYM theories in the harmonic superspace were formulated in [19, 20], the background field method for these theories was introduced in [121–123], and examples of quantum calculations are given in [111, 120, 124–126]. The most important results presented in these papers are obtaining of the holomorphic action of $\mathcal{N} = 2$ matter hypermultiplets in the external $\mathcal{N} = 2$ gauge superfield, the calculation of the one-loop nonholomorphic effective potential in the $\mathcal{N} = 4$ SYM theory and proof of its absence in higher loops, finding of the one-loop effective action for $\mathcal{N} = 4$ SYM theory for a constant strength tensor F_{ab}, and computing of a superconformal anomaly of $\mathcal{N} = 2$ matter interacting with $\mathcal{N} = 2$ supergravity. During last years, other important results of these investigations were the calculations of two-loop contributions depending on derivatives of $\mathcal{N} = 2$ SYM strength $\mathcal{W}$ [127, 128] and of the one-loop effective action in the matter hypermultiplet sector [129], and the development of a quantum approach for $\mathcal{N} = 3$ SYM theory [130] (the $\mathcal{N} = 3$ harmonic superspace technique was introduced in the paper [60]). Also, it is necessary to mention the intensive studies of three-dimensional extended supersymmetric theories, especially $\mathcal{N} = 6$ and $\mathcal{N} = 8$ Chern-Simons theories (see e.g. [4]), where also the harmonic superspace approach has been successfully developed and applied (see e.g. [131]).

2. *Noncommutative supersymmetric theories.* Noncommutative theories have been intensively studied during recent years. The concept of the space-time noncommutativity was introduced to quantum field theory being motivated by some consequences of the D-branes theory [132] and by the interest to behaviour of quantum theories at very small distances where quantum fluctuations of geometry are essential. The consideration of supersymmetric noncommutative theories is quite natural. During last years some interesting results in studying of noncommutative supersymmetric theories were obtained but they were mostly based on component approach. The first superfield results were the calculation of leading ($\sim F^4$) correction to the one-loop effective action for $\mathcal{N} = 4$ SYM theory [108] and the formulation of a supergraph technique for the noncommutative Wess-Zumino model [29]. Further, the quantum superfield studies for noncommutative extensions of the Wess-Zumino model [133] and four-dimensional superfield QED [103] and SYM theories [104] were carried out. There are also various examples of calculations in three-dimensional supersymmetric field theories presented in [15, 54], and some of them have been considered in this book. These theories were shown to be consistent in the sense of the absence of the nonintegrable UV/IR infrared divergences. Thus, we can speak about constructing of consistent noncommutative generalizations of the supersymmetric theories of electromagnetic, strong and weak interactions. Therefore, the next most important problem could consist in the development of the noncommu-

tative supersymmetric generalization for the last fundamental interaction—the gravitational one. However, this problem is certainly extremely difficult (see discussion of the problem e.g. in [134, 135]).

3. *Noncommutative superspace* One more approach in the superfield quantum theory is based on using of the noncommutative superspace [31]. Within it, the fermionic superspace coordinates form the Clifford algebra instead of the Grassmann one, which, in particular, leads to the modified construction of the Moyal product. Within the framework of this approach, the generalizations of the Wess-Zumino (see e.g. [136] and references therein), gauge (see e.g. [137] and references therein) and general chiral superfield models [138] were studied. A three-dimensional version of the noncommutative superspace [55] has been presented in the Sect. 3.7 of this review.

4. *Lorentz-breaking supersymmetric theories.* As it is known (see e.g. [139]), the breaking of the Lorentz symmetry is introduced through additive terms proportional to constant vectors, or, in general, constant tensors, which introduce privileged directions in space-time. Therefore, the natural question is—how one can implement Lorentz symmetry breaking in supersymmetric theories? There are three known manners to do it. Within one of them, we introduce the Lorentz-breaking operator at the level of the superfield action, without modifying structures of superfields or a supersymmetry algebra. This approach, motivated by the idea to construct a supersymmetric generalization of Horava-Lifshitz-like theories displaying a space-time anisotropy, is presented in [140]. Within another manner, one introduces a new superfield so that some its components are proportional to Lorentz-breaking vectors (tensors) which allows to construct supersymmetric extensions of known Lorentz-breaking terms, such as the Carroll-Field-Jackiw term and the aether term [141]. Finally, the third manner is based on the Kostelecky-Berger deformation of the supersymmetry algebra through the replacement $\partial_m \to \partial_m + k_{mn}\partial^n$ within supersymmetry generators (4.11), with k_{mn} are constant tensors defined in such a manner that $|k_{mn}| \ll 1$ for any m, n to ensure smallness of the Lorentz symmetry breaking. It was explicitly demonstrated that in superfield theories based on this approach, all perturbative superfield calculations can be performed with no more difficulties that in usual superfield theories, see e.g. [143].

Besides of all this, there are a lot of applications of the superfield approach to various problems of supersymmetric quantum field theory, e.g. to studying of AdS/CFT correspondence which was carried out mostly on the base of a component approach, and of course to considering of many problems originated from strings/M-theory.

As a final conclusion, we can suppose that superfield approach in quantum field theory is a very perspective methodology, and there are a lot of ways for its development and more applications, including contexts of phenomenology and even condensed matter.

Bibliography

1. M. Green, J. Schwarz, E. Witten. *Theory of Superstrings*, vol. 1–2 (Cambridge University Press, Cambridge, UK, 1987)
2. S. Kovacs, Int. J. Mod. Phys. A **21**, 4555 (2006), hep-th/9902047
3. Z. Bern, J.J. Carrasco, L. Dixon, H. Johansson, D.A. Kosower, R. Roiban, Phys. Rev. Lett. **98**, 161303 (2007), hep-th/0702112
4. O. Aharony, O. Bergmann, D.L. Jafferis, J. Maldacena, JHEP **0810**, 091 (2008), arXiv: 0806.1218
5. N. Lambert, Ann. Rev. Nucl. Part. Sci. **62**, 285 (2012), arXiv:1203.4244
6. D.V. Volkov, V.P. Akulov, JETP Lett. **16**, 621 (1972); Phys. Lett. **B46**, 109 (1973)
7. Y.A. Golfand, E.S. Lichtman, JETP Lett. **13**, 323 (1971)
8. J. Wess, B. Zumino, Nucl. Phys. B **70**, 139 (1974)
9. J. Wess, B. Zumino, Nucl. Phys. B **78**, 1 (1974)
10. G. Kane, M. Shifman (ed.), *The Supersymmetric World* (World Scientific, 2000)
11. J. Wess, B. Zumino, Phys. Lett. B **49**, 52 (1974)
12. A. Salam, J. Strathdee, Nucl. Phys. B **76**, 477 (1974)
13. S. Ferrara, J. Wess, B. Zumino, Phys. Lett. B **51**, 239 (1974)
14. P.S. Howe, K.S. Stelle, P.K. Townsend, Nucl. Phys. B **236**, 125 (1984)
15. A.F. Ferrari, H.O. Girotti, M. Gomes, A.Yu. Petrov, A.A. Ribeiro, A.J. da Silva, Phys. Lett. B **577**, 83 (2003), hep-th/0309193; E.A. Asano, H.O. Girotti, M. Gomes, A.Yu. Petrov, A.G. Rodrigues, A.J. da Silva, Phys. Rev. D **69**, 105012 (2004), hep-th/0402013; E.A. Asano, L.C.T. Brito, M. Gomes, A.Yu. Petrov, A.J. da Silva, Phys. Rev. **D71**, 105005 (2005), hep-th/0410257; A.F. Ferrari, M. Gomes, J.R. Nascimento, A.Yu. Petrov, A.J. da Silva, E.O. Silva, Phys. Rev. **D77**, 025002 (2008), arXiv: 0708.1002
16. L. Brink, O. Lindgren, B.E.W. Nilsson, Phys. Lett. B **123**, 323 (1983)
17. O. Piguet, K. Sibold, Phys. Lett. **B177**, 373 (1986); Int. J. Mod. Phys. **A1**, 913 (1986); C. Lucchesi, O. Piguet, K. Sibold, Phys. Lett. **B201**, 241 (1988); Helv. Phys. Acta **61**, 321 (1988)
18. J. Wess, B. Zumino, Phys. Lett. **B66**, 361 (1977); **B74**, 51 (1978)
19. A. Galperin, E. Ivanov, S. Kalitzin, V. Ogievetsky, E. Sokachev, Class. Quant. Grav. **1**, 469 (1984)
20. A. Galperin, E. Ivanov, V. Ogievetsky, E. Sokachev, Class. Quant. Grav. **2**, 601 (1985); **2**, 617 (1985)
21. M.B. Green, J.H. Schwarz, Phys. Lett. B **136**, 367 (1984)
22. W. Siegel, Phys. Lett. B **80**, 220 (1979)

© The Editor(s) (if applicable) and The Author(s), under exclusive license to
Springer Nature Switzerland AG 2021
A. Petrov, *Quantum Superfield Supersymmetry*, Fundamental Theories of Physics 202,
https://doi.org/10.1007/978-3-030-68136-4

23. S.J. Gates, M.T. Grisaru, M. Rocek, W. Siegel, *Superspace or One Thousand and One Lectures in Supersymmetry* (Benjamin/Cummings, 1983)
24. E.M.C. Abreu, M.A. De Andrade, L.P.G. De Assis, J.A. Helayel-Neto, A.L.M.A. Nogueira, R.C. Paschoal, JHEP **1105**, 001 (2011), arXiv:1002.2660 [hep-th]
25. I. Jack, D.R.T. Jones, P. West, Phys. Lett. B **258**, 382 (1991)
26. P. West, Phys. Lett. B **261**, 396 (1991)
27. I.L. Buchbinder, S.M. Kuzenko, J.V. Yarevskaya, Nucl. Phys. **B411**, 665 (1994); Yad. Fiz. (Phys. Atom. Nucl.) **56**, 5, 202 (1993)
28. N. Seiberg, E. Witten, Nucl. Phys. B **426**, 19 (1994), hep-th/9407087
29. A.A. Bichl, J.M. Grimstrup, H. Grosse, L. Popp, M. Schweda, R. Wulkenhaar, JHEP **10**, 046 (2000), hep-th/0007050
30. S. Minwalla, M. van Raamsdonk, N. Seiberg, JHEP **0002**, 020 (2000), hep-th/9912072
31. N. Seiberg, JHEP **0306**, 010 (2003), hep-th/0305248
32. J. Wess, J. Bagger, *Supersymmetry and supergravity* (Princeton University Press, Princeton, 1983)
33. I.L. Buchbinder, S.M. Kuzenko, *Ideas and Methods of Supersymmetry and Supergravity* (IOP Publishing, Bristol and Philadelphia, 1998)
34. S. Coleman, S. Weinberg, Phys. Rev. D **7**, 1888 (1973)
35. I.L. Buchbinder, S.D. Odintsov, I.L. Shapiro, *Effective Action in Quantum Gravity* (IOP Publishing, Bristol and Philadelphia, 1992)
36. B.S. De Witt, *Dynamical Theory of Groups and Fields* (Princeton University Press, Princeton, 1987)
37. F. Ruiz Ruiz, P. van Nieuwenhuizen. Lectures on Supersymmetry and Supergravity in 2+1 Dimensions and Regularization of Supersymmetric Theories. Published in: Recent developments in gravitation and mathematical physics, Proceedings, 2nd Mexican School, Tlaxcala, Mexico, 1996
38. F.A. Berezin, *Introduction to Superalgebra* (Springer, 1997)
39. W. Kummer, M. Schweda, Phys. Lett. **B141**, 363 (1984); W. Kummer, H. Mistelberger, P. Schaller, M. Schweda, T. Kreuzberger, Nucl. Phys. **B281**, 411 (1987); W. Kummer, H. Mistelberger, P. Schaller, M. Schweda, Z. Phys. **C40**, 91 (1988); W. Kummer, H. Mistelberger, P. Schaller, Z. Phys. **C40**, 103 (1988)
40. A.F. Ferrari, M. Gomes, A.C. Lehum, J.R. Nascimento, A.Yu. Petrov, A.J. da Silva, Phys. Lett. B **678**, 233 (2009), arXiv: 0812.3134
41. A.F. Ferrari, M. Gomes, A.C. Lehum, A.Yu. Petrov, A.J. da Silva, Phys. Rev. D **77**, 065005 (2008), arXiv: 0709.3501
42. H.O. Girotti, M. Gomes, A.Yu. Petrov, V.O. Rivelles, A.J. da Silva, Phys. Lett. B **521**, 119 (2001), hep-th/0109222
43. P. West, *Introduction to Supersymmetry and Supergravity* (World Scientific, 1989)
44. T. Inami, Y. Saito, M. Yamamoto, Progr. Theor. Phys. **103**, 1283 (2000), hep-th/0003013; M. Ciuchini, J.A. Gracey, Nucl. Phys. **B454**, 103 (1995), hep-th/9508176; J.H. Cho, S.O. Hahn, P. Oh, C. Park, J.H. Park, JHEP **01**, 057 (2004), hep-th/0312088
45. A.F. Ferrari, M. Gomes, A.C. Lehum, J.R. Nascimento, A.Yu. Petrov, A.J. da Silva, Phys. Lett. B **678**, 500 (2009), arXiv: 0901.0679
46. C.P. Burgess, Nucl. Phys. B **216**, 459 (1983)
47. S. Ojima, Progr. Theor. Phys. **81**, 512 (1989)
48. T. Mariz, J.R. Nascimento, A.Yu. Petrov, L.Y. Santos, A.J. da Silva, Phys. Lett. B **661**, 312 (2008), arXiv: 0708.3348
49. F.S. Gama, J.R. Nascimento, A.Yu. Petrov, Phys. Rev. D **88**, 045021 (2013), arXiv:1307.3190
50. H. Snyder, Phys. Rev. **71**, 38 (1947)
51. T. Filk, Phys. Lett. B **376**, 53 (1996)
52. I.Ya. Aref'eva, D.M. Belov, A.S. Koshelev, Phys. Lett. B **476**, 431 (2000), hep-th/9912075
53. L. Alvarez-Gaume, J.L.F. Barbon, R. Zwicky, JHEP **0105**, 057 (2001), hep-th/0103069
54. H.O. Girotti, M. Gomes, V.O. Rivelles, A.J. da Silva, Int. J. Mod. Phys. A **17**, 1503 (2002), hep-th/0102101

55. A.F. Ferrari, M. Gomes, J.R. Nascimento, A.Yu. Petrov, A.J. da Silva, Phys. Rev. D **74**, 12501 (2006), hep-th/0607087

56. F.S. Gama, J.R. Nascimento, A.Yu. Petrov, Int. J. Mod. Phys. A **31**, 1650055 (2016), arXiv: 1406.5418

57. I.L. Buchbinder, N.G. Pletnev, I.B. Samsonov, JHEP **1004**, 124 (2010), arXiv: 1003.4806; JHEP **1101**, 121 (2011), arXiv: 1010.4967; I.L. Buchbinder, N.G. Pletnev, JHEP **1111**, 085 (2011), arXiv: 1108.2966

58. D. Gaiotto, X. Yin, JHEP **0708**, 056 (2007), arXiv: 0704.3740

59. U. Lindstrom, M. Rocek, Commun. Math. Phys. **128**, 191 (1990); F. Gonzalez-Rey, M. Wiles, U. Lindstrom, R. von Unge, Nucl. Phys. **B516**, 426 (1998), hep-th/9710250

60. F. Delduc, J. McCabe, Class. Quant. Grav. **6**, 233 (1989)

61. V. Ogievetsky, L. Mezinchesku, Uspehi Fiz. Nauk (Phys. Sci. Achievments) **117**, 637 (1975) (in Russian)

62. I.L. Buchbinder, A.Yu. Petrov, Yad. Fiz. (Phys. Atom. Nucl.) **63**, 9, 2557 (2000)

63. I.L. Buchbinder, A. Yu. Petrov, Class. Quant. Grav. **13**, 2081 (1996), hep-th/9511205

64. T. Fujimori, M. Nitta, Y. Yamada, JHEP **1609**, 106 (2016), arXiv:1608.01843 [hep-th]

65. S.M. Kuzenko, S.J. Tyler, JHEP **1104**, 057 (2011), arXiv:1102.3042 [hep-th]

66. M.T. Grisaru, M. Rocek, W. Siegel, Nucl. Phys. B **159**, 429 (1979)

67. P. West, Phys. Lett. B **258**, 375 (1991)

68. H.O. Girotti, M. Gomes, A.Yu. Petrov, V.O. Rivelles, A.J. da Silva, Phys. Rev. D **67**, 125003 (2003), hep-th/0207220

69. I. Jack, D.R.T. Jones, Adv. Ser. Direct. High Energy Phys. **21**, 494 (2010), hep-ph/9707278

70. S. Groot Nibbelink, T.S. Nyawelo, JHEP **0601**, 034 (2006), hep-th/0511004

71. J. Collins, *Renormalization* (University Press, UK, 1984)

72. I.L. Buchbinder, S.M. Kuzenko, A.Yu. Petrov, Phys. Lett. B **321**, 372 (1994)

73. I.L. Buchbinder, S.M. Kuzenko, A.Yu. Petrov, Yad. Fiz. (Phys. Atom. Nucl.) **59**, 1, 157 (1996)

74. I.N. McArthur, T.D. Gargett, Nucl. Phys. B **497**, 525 (1997), hep-th/9705200

75. A. Pickering, P. West, Phys. Lett. B **353**, 54 (1997), hep-th/9604147

76. S.M. Kuzenko, S.J. Tyler, JHEP **1409**, 135 (2014), arXiv:1407.5270 [hep-th]

77. N.I. Usyukina, Theor. Math. Phys. **88**, 683 (1991); N.I. Usyukina, A. Davydychev, Phys. Lett. **B298**, 363 (1993)

78. S.M. Kuzenko, J.V. Yarevskaya, Yad. Fiz. (Phys. Atom. Nucl.) **56**, 5, 195 (1993)

79. K. Wilson, J. Kogut, Renormalization group and ϵ expansion. Phys. Rep. **12**, 75 (1974)

80. G. Cleaver, M. Cvetič, J.R. Espinosa, L. Everett, P. Langacker, Nucl. Phys. B **525**, 3 (1998), hep-th/9711178; G. Cleaver, M. Cvetič, J.R. Espinosa, L. Everett, P. Langacker, J. Wang, Phys. Rev, **D59**, 115003 (1999), hep-th/9811355; M. Cvetič, L. Everett, J. Wang, Nucl. Phys. **B538**, 52 (1999), hep-ph/9807321

81. H. Georgi, Ann. Rev. Nucl. Part. Sci. **43**, 209 (1993)

82. A.Yu. Petrov, Effective action in general chiral superfield model, hep-th/0002013

83. I.L. Buchbinder, A.Yu. Petrov, Phys. Lett. B **461**, 209 (1999), hep-th/9905062

84. K. Symanzik. Commun. Math. Phys. **34**, 7 (1973); T. Appelquist, J. Carrazone, Phys. Rev. **D11**, 2856 (1975)

85. I.L. Buchbinder, M. Cvetič, A.Yu. Petrov, Mod. Phys. Lett. A **15**, 783 (2000). hep-th/9903243; Nucl. Phys. **B571**, 358 (2000), hep-th/9906141

86. J. Iliopoulos, B. Zumino, Nucl. Phys. B **76**, 310 (1974)

87. K. Stelle, Phys. Rev. D **16**, 953 (1977)

88. I. Antoniadis, E. Mottola, Phys. Rev. D **45**, 2013 (1992)

89. I.L. Buchbinder, S.M. Kuzenko, Phys. Lett. B **202**, 233 (1988)

90. I.L. Buchbinder, A.Yu. Petrov, Class. Quant. Grav. **14**, 21 (1997), hep-th/9607217

91. A.D. Dolgov, M. Kawasaki, Phys. Lett. **B573**, 1 (2003), astro-ph/0307285; S. Nojiri, S.D. Odintsov, Phys. Rev. **D68**, 123512 (2003), hep-th/0307288

92. P. Horava, Phys. Rev. D **79**, 084008 (2009), arXiv: 0901.3775

93. I. Antoniadis, E. Dudas, D. Ghilencea, JHEP **03**, 045 (2008), arXiv: 0708.0383

94. A.V. Smilga, Nucl. Phys. B **706**, 598 (2005), hep-th/0407231

95. I. Antoniadis, E. Dudas, D.M. Ghilencea, Nucl. Phys. B **767**, 29 (2007), hep-th/0608094
96. M. Gomes, J.R. Nascimento, A.Yu. Petrov, A.J. da Silva, Phys. Lett. B **682**, 229 (2009), arXiv: 0908.0900
97. M. Abramowitz, I.A. Stegun, *Handbook of Mathematical Functions* (National Bureau of Standards, Washington, D.C., 1964)
98. F.S. Gama, M. Gomes, J.R. Nascimento, A.Yu. Petrov, A.J. da Silva, Phys. Rev. D **84**, 045001 (2011), arXiv: 1101.0724
99. M. Cvetič, T. Mariz, A.Yu. Petrov, Phys. Rev. D **92**, 085041 (2015), arXiv:1509.00251 [hep-th]
100. M.T. Grisaru, M. Rocek, R. von Unge, Phys. Lett. B **383**, 415 (1996), hep-th/9605149
101. A. De Giovanni, M.T. Grisaru, D. Zanon, Phys. Lett. B **409**, 251 (1997), hep-th/9706013
102. P.S. Howe, K. Stelle, P. West, Phys. Lett. B **124**, 55 (1983)
103. A.F. Ferrari, H.O. Girotti, M. Gomes, A.Yu. Petrov, A.A. Ribeiro, V.O. Rivelles, A.J. da Silva, Phys. Rev. D **69**, 025008 (2004), hep-th/0309154
104. A.F. Ferrari, H.O. Girotti, M. Gomes, A.Yu. Petrov, A.A. Ribeiro, V.O. Rivelles, A.J. da Silva, Phys. Rev. D **70**, 085012 (2004), hep-th/0407040
105. M.T. Grisaru, W. Siegel, Nucl. Phys. B **201**, 292 (1982)
106. M.T. Grisaru, D. Zanon, Phys. Lett. **B142**, 359 (1984); Nucl. Phys. **B252**, 578 (1985)
107. A.A. Bichl, M. Ertl, A. Gerhold, J.M. Grimstrup, H. Grosse, L. Popp, V. Putz, M. Schweda, R. Wulkenhaar, Int. J. Mod. Phys. A **19**, 4231 (2004), hep-th/0203141
108. D. Zanon, Phys. Lett. B **502**, 265 (2001), hep-th/0012009; Phys. Lett. **B504**, 101 (2001), hep-th/0009196; A. Santambrogio, D. Zanon, JHEP **0101**, 024 (2001), hep-th/0010275; M. Pernici, A. Santambrogio, D. Zanon, Phys. Lett. **B504**, 131 (2001), hep-th/0011140
109. E. Elizalde, *Zeta Regularization Techniques with Applications* (World Scientific, 1994)
110. N.G. Pletnev, A.T. Banin, Phys. Rev. D **60**, 105017 (1999), hep-th/9811031; I.L. Buchbinder, S.M. Kuzenko, A.A. Tseytlin, Phys. Rev. **D62**, 045001 (2000), hep-th/9911221
111. I.L. Buchbinder, S.M. Kuzenko, Phys. Lett. B **446**, 216 (1999), hep-th/9810239; Mod. Phys. Lett. **A13**, 1623 (1998), hep-th/9804168
112. F.S. Gama, J.R. Nascimento, A.Yu. Petrov, Phys. Rev. D **101**, 105018 (2020), arXiv:2004.09299 [hep-th]
113. F.S. Gama, M. Gomes, J.R. Nascimento, A.Yu. Petrov, A.J. da Silva, Phys. Rev. **D91**, 065038 (2015) Erratum: [Phys. Rev. **D91**, 129901 (2015)], arXiv:1501.04061 [hep-th], C.A.S. Almeida, F.S. Gama, R.V. Maluf, J.R. Nascimento, A.Yu. Petrov, Phys. Rev. **D92**, 085003 (2015), arXiv:1506.04001 [hep-th]
114. M. Luty, TASI lectures on supersymmetry breaking, hep-th/0509029 (2004)
115. K. Intriligator, N. Seiberg, Class. Quant. Grav. **24**, S741 (2007), hep-ph/0702069
116. P. Nath, R. Arnowitt, Phys. Rev. D **56**, 2820 (1997), hep-ph/9701301
117. P. Fayet, J. Iliopoulos, Phys. Lett. B **51**, 461 (1974)
118. L. O'Raifeartaigh, Nucl. Phys. B **96**, 331 (1975)
119. D. Shih, JHEP **0802**, 091 (2008), hep-th/0703196
120. I.L. Buchbinder, A.Yu. Petrov, Phys. Lett. B **469**, 482 (2000), hep-th/0003265
121. E.I. Buchbinder, I.L. Buchbinder, E.A. Ivanov, S.M. Kuzenko, B.A. Ovrut, Phys. Lett. B **412**, 309 (1997), hep-th/9703147
122. E.I. Buchbinder, I.L. Buchbinder, S.M. Kuzenko, B.A. Ovrut, Phys. Lett. B **417**, 61 (1997), hep-th/9704214
123. I.L. Buchbinder, S.M. Kuzenko, B.A. Ovrut, Phys. Lett. B **433**, 335 (1997), hep-th/9710142
124. I.L. Buchbinder, I.B. Samsonov, Mod. Phys. Lett. A **14**, 2537 (1999), hep-th/9909183
125. S.M. Kuzenko, I.N. McArthur, Phys. Lett. B **506**, 140 (2001), hep-th/0101127
126. S.M. Kuzenko, S. Theisen, Class. Quant. Grav. **17**, 665 (2000), hep-th/9907107
127. I.L. Buchbinder, A.Yu. Petrov, A.A. Tseytlin, Nucl. Phys. B **621**, 179 (2002), hep-th/0110173
128. S.M. Kuzenko, I.N. McArthur, JHEP **0310**, 029 (2003), hep-th/0308136; Nucl. Phys. B683, 3 (2004), hep-th/0310025; Nucl. Phys. **B697**, 89 (2004), hep-th/0403240
129. I.L. Buchbinder, E.A. Ivanov, A. Yu. Petrov, Nucl. Phys. B **653**, 64 (2003), hep-th/0210241
130. I.L. Buchbinder, E.A. Ivanov, I.B. Samsonov, B.M. Zupnik, Nucl. Phys. B **689**, 91 (2004), hep-th/0403053

131. I.L. Buchbnder, E.A. Ivanov, O. Lechtenfeld, N.G. Pletnev, I.B. Samsonov, B.M. Zupnik, JHEP **0903**, 096 (2009), arXiv: 0811.4774; JHEP **0910**, 075 (2009), arXiv: 0909.2970
132. N. Seiberg, E. Witten, JHEP **9909**, 032 (1999), hep-th/9908142
133. I.L. Buchbinder, M. Gomes, A.Yu. Petrov, V.O. Rivelles, Phys. Lett. B **517**, 191 (2001), hep-th/0107022
134. A.H. Chamseddine, J. Math. Phys. **44**, 2534 (2003), hep-th/0202137
135. M.A. Cardella, D. Zanon, Class. Quant. Grav. **20**, L95 (2003), hep-th/0212071
136. M.T. Grisaru, S. Penati, A. Romagnoni, Class. Quant. Grav. **S1391** (2004), hep-th/0401174; A.T. Banin, I.L. Buchbinder, N.G. Pletnev, JHEP **0407**, 011 (2004), hep-th/0405063
137. S. Penati, A. Romagnoni, JHEP **0502**, 064 (2005), hep-th/0412041
138. O.D. Azorkina, A.T. Banin, I.L. Buchbinder, N.G. Pletnev, Mod. Phys. Lett. A **20**, 1423 (2005), hep-th/0502008
139. D. Colladay, V.A. Kostelecky, Phys. Rev. **D55**, 6760 (1997), hep-ph/9703464; D. Colladay, V.A. Kostelecky, Phys. Rev. **D58**, 116002 (1998), hep-ph/9809521
140. M. Gomes, J.R. Nascimento, A.Yu. Petrov, A.J. da Silva, Phys. Rev. D **90**, 125022 (2014), arXiv:1408.6499 [hep-th]
141. H. Belich, J.L. Boldo, L.P. Colatto, J.A. Helayel-Neto, A.L.M.A. Nogueira, Phys. Rev. D **68**, 065030 (2003), [hep-th/0304166]; H. Belich, L.D. Bernald, P. Gaete, J.A. Helayel-Neto, F.J.L. Leal, Eur. Phys. J. **C75**, 291 (2015), arXiv:1502.06126
142. M.S. Berger, V.A. Kostelecky, Phys. Rev. D **65**, 091701 (2002), hep-th/0112243
143. C.F. Farias, A.C. Lehum, J.R. Nascimento, A.Yu. Petrov, Phys. Rev. D **86**, 065035 (2012), arXiv:1206.4508 [hep-th]

Index

© The Editor(s) (if applicable) and The Author(s), under exclusive license to
Springer Nature Switzerland AG 2021
A. Petrov, *Quantum Superfield Supersymmetry*, Fundamental Theories of Physics 202,
https://doi.org/10.1007/978-3-030-68136-4